S0-EIH-067

# THE BRITISH ECONOMY

# THE BRITISH ECONOMY

## VOLUME 1
## The Years of Turmoil 1920-1951

**Derek H Aldcroft**

*Professor of Economic History*
*University of Leicester*

HUMANITIES PRESS INTERNATIONAL, INC.
ATLANTIC HIGHLANDS, N.J.

First published in 1986 in the United States of America by
Humanities Press International, Inc., Atlantic Highlands, NJ 07716.



**Library of Congress Cataloging-in-Publication Data**

Aldcroft, Derek Howard.
The British economy.

Bibliography: p
Contents: v. 1 Years of Turmoil, 1920-1951.
1 Great Britain — Economic policy — 1918-1945.
2 Great Britain — Economic policy — 1945- .
3 Great Britain — Economic conditions — 1918-1945.
4 Great Britain — Economic conditions — 1945-
I Title.
HC256.3.A584 1986 338.941 85-27052
ISBN 0-391-03379-4 (v. 1)
ISBN 0-391-03380-8 (pbk. : v. 1)



PRINTED IN GREAT BRITAIN

# Contents

# Preface

This is the first volume of what is intended to be an extended study of the British economy from 1920 to the present day. Energy permitting, there promise to be two more volumes, one dealing with Britain's growth failure in the super-growth era through to the early 1970s, and a final one on the search for stability in the contemporary world. It seemed appropriate to terminate the first volume with the fall of the Labour Government in 1951 since by that time the reconstruction phase was more or less complete and a new vista of events and circumstances was about to emerge. A second reason for stopping at that point was the fact that the volume was already long enough.

I have not attempted to give a detailed account of every aspect of economic life in Britain during this thirty-year period, since such an exercise would stimulate neither writer nor reader. What I have done is to take a close look at some of the major economic issues and problems of the period, to debate these in the light of the many new and stimulating contributions made in recent years, some of which draw upon contemporary economic analysis, and to pay particular attention to the role of economic policy, which has been the subject of considerable controversy in the recent literature. Many of the interwar problems in particular are relevant to today's world and where possible I have made allusion to that fact. I should like to think that a study of the past can help to provide a better understanding of our current difficulties.

I am very grateful to Mrs Margaret Christie and Mrs Gillian Austen for the preparation of the typescript, and additionally, I am indebted to Mrs Austen once again for her skill in spotting errors, often my spelling ones!

Derek H. Aldcroft
University of Leicester

# 1 Chequered Decade: the 1920s

If from a social point of view the 1920s had much to offer in terms of a release from many of the constraints of Victorian conventions, the decade was something of a disappointment economically. It was a turbulent and eventful decade not unlike that to follow half a century later. And whereas the 1930s could at least show a remarkably strong and sustained recovery from the depression of 1929-32 growth in the 1920s was slow and patchy, and Britain certainly did not share fully in the strong boom enjoyed by the United States at this time. Moreover, by the end of the decade it became all too clear that this country was suffering from several long-term intractable problems, namely high unemployment, industrial and regional imbalance and stagnating exports, all of which were closely interrelated.

## A Decade of Instability

Britain's poor record of achievement in the 1920s is perhaps the more remarkable given the fact that she emerged from the war in a stronger and healthier state, relatively speaking, than her European neighbours. True, Britain's economic strength could not match that of her wartime ally, the United States, which had emerged as the main beneficiary of the war. On the other hand, while much was made at the time of the loss of shipping, sales of overseas assets and the lost generation of Britons, the actual destruction and dislocation caused by the war were minimal compared with what happened in many European countries. Moreover, despite the strengthened power of labour and the social and political unrest immediately following the war, Britain appeared positively tranquil by comparison with the convulsive state of society in countries such as Poland, Austria, Germany and even Belgium and France, where living standards and social conditions had deteriorated badly during the course of the war.

In fact initially the outlook for the British economy seemed to be full of promise. Output had not collapsed during the course of the war; if anything it had increased slightly though in a direction away from peacetime needs. Under the exigencies of wartime demands many manufacturers had been forced to abandon their 'cottage industry' practices and streamline their methods of production to improve efficiency. The government too, in the latter part of the war, had laid grandiose reconstruction schemes designed to provide a better society and to reap the benefits from the experience of war. Furthermore, the transition from war — to peacetime production, the demobilisation of the armed forces, and the renewal of equipment after several years of neglect, not to mention the highly liquid state of the economy and the prospects of recapturing lost markets abroad — appeared to offer businessmen lucrative opportunities once they were released from the shackles of wartime controls.

Conditions were in fact tailor-made for a boom in activity soon after the termination of hostilities. As a result of the forced neglect of maintenance and renewal during the war, there was a considerable backlog of investment to be made good. Worn-out machinery and plant required replacing especially in those industries, notably engineering, that had been geared to military production. Some of the large basic industries under government control, such as mining and the railways, had been starved of investment resources during the war. Housing too had been badly neglected. In fact few new houses were built during the period, so that by the end of the war there was a serious housing shortage estimated to be in the region of one million units. There was also a serious scarcity of consumer products as a result of labour and materials shortages and the shift of capacity to wartime production. Many industries had reduced their output for the domestic market by one half or more and by 1918 the number of workers directly producing goods for the civilian market was about one third that of prewar years. As a result private consumers' expenditure fell by some 20 per cent between 1913 and 1918. Finally, the demand for exports was expected to rise sharply from the low level reached in 1918.

The deficiency on the supply side occurred at a time when there was a large pent-up demand for goods. Excess demand at the end of the war was probably in the region of 10-15 per cent. This situation arose because of the highly liquid state of the economy. Because of restrictions on investment and consumption for normal requirements during the war, business, the banks and the public in general had accumulated a considerable amount of cash or near liquid assets, for example government bonds. The methods of wartime finance, in particular extensive borrowing through the large scale issue of Treasury Bills, was instrumental in increasing the liquid state of the economy.

Thus, after a temporary lull in economic activity following the Armistice in November, a boom developed in the spring of 1919. It was considerably shorter than originally anticipated and it had some unsavoury and frothy characteristics. Needless to say it ended in disaster. What seems to have happened is that frustrated consumers unleashed their pent-up purchasing power on a market which could not fully meet their needs because of the dislocation involved in converting production from a war to a peacetime basis. Consumers' expenditure rose by no less than 21 per cent between 1918-19, a very large increase in a single year, and it occurred at a time when government spending was still running at a fairly high level. Furthermore, in the following year there was also a sharp rise in exports. Despite frantic efforts by businessmen to cope with the situation and repair the backlog of investment, productive capacity was insufficient in the short term to meet the sudden increase in demand, despite a significant rise in output through to 1920. Hence the price effects came through more strongly than those of output and these were aggravated by a significant rise in money wages as the unions sought to cash in on their new-found strength and freedom from wartime agreements. Both the price-cost spiral and in some cases the supply position were aggravated by several factors including the sudden removal of most wartime controls in the spring of 1919, shortages of labour and raw materials, a wave of strikes, transport bottlenecks, the depreciation of sterling and initially by the government's lax fiscal and monetary policies.

The boom lasted only just over a year, March 1919 to April 1920, and in many respects it was a very artificial and speculative one. The upswing was dominated by inflationary pressures largely because production could not keep pace with effective demand. Its most outstanding feature was the extensive speculation carried on in commodities, securities and real estate and the vast number of industrial transactions and flotations secured at highly inflated prices. The financial orgy was underpinned by the very liquid state of many firms following the realisation of big profit gains during the war, the large-scale creation of bank credit and easy money conditions. Much speculative activity was based on borrowed money. Almost £400 million of bank credit was made available for industrial and commercial purposes in the period January 1919 to April 1920, while total bank clearings rose by nearly 83 per cent between 1918-20. On the London money market new capital issues peaked at £384.2 million in 1920, a figure not again surpassed until the 1950s, as against £65.3 million in 1918. Most of the increase in new issues was for domestic purposes, as opposed to foreign account before the war.

Perhaps the most unhappy feature of the whole affair was that the worst speculative excesses occurred in many of the older industries which, in view of their poor future growth potential, could ill-afford to engage in such frivolous activities. Most noteworthy in this respect were the shipping, shipbuilding, textile (especially cotton) and engineering industries, where the flotation of new companies, the sale of old ones and the issue of new shares became almost a daily event in 1919. The anticipation of high profits and capital gain lured spectulators into risky ventures as a result of which many companies were bought up and refloated at inflated capital values often with the connivance of the banks. The cotton industry provides a good illustration, where 109 mills with an original share capital of £4.5 million were sold for no less than £31.7 million. In all, some 42 per cent of cotton spinning capacity is said to have changed hands between 1919-20 at an estimated seven times prewar values. Ships became an even greater counter for the speculator. Because of a temporary shortage of cargo capacity ship prices rocketed and, as the pages of the shipping journals testify, many vessels changed hands several times at ever-inflated prices. The price of *Fairplay*'s new ready cargo steamer of 7,500 tons rose from £169,000 to £259,000 at the peak in March 1920, compared with a prewar price of £60,000, but by the end of 1920 the price had collapsed to £105,000 which illustrates the degree of speculative froth. Second-hand tonnage prices were even more buoyant; in some cases vessels appreciated by 100 per cent or more within a matter of months. The *Annandale*, for example, sold for £10,000 in January 1920 and was resold in May for £25,000, a profit of 150 per cent.

Such speculative excesses were by no means isolated examples, as a glance at the contemporary press will show. What is important to note is that they inevitably had disastrous consequences for many of the older industries in view of the subsequent course of market events. Not only were their accumulated profits dissipated in unsound assets, but they were left with a heavy burden of debt as a result of increased interest liabilities, the issue of bonus shares and the watering of capital stock at a time when the earning power of the acquired assets was soon to be much diminished. The cost of overcapitalisation was to remain a millstone round the necks of many firms throughout the interwar period, the agonies of which were prolonged for a time by the generosity of the banks who, in an effort to protect their advances, kept many firms afloat, especially in the cotton industry. Moreover, in some cases, the shipbuilding industry in particular, the boom brought into existence sufficient capacity to last a decade or more. The number of shipbuilding berths had risen to 806 by 1920 compared with 580 in 1914. And often this capacity, like that brought into existence in the subsequent German inflation, was not particularly suited to future needs.

The boom collapsed almost as sharply as it had begun. The end was formally signalised in April 1920 when Bank rate was hoisted to 7 per cent. However, worries about the inflationary potential getting out of control and concern about the need to restrain the high level of public spending and reduce the large floating debt had prompted the government to reverse course before the spring of 1920. The heavy fiscal deficit of the previous year, which had helped to fuel the boom initially, was converted to a surplus by the final quarter of 1919, by which time interest rates were rising and the monetary aggregates were being slowed down. This increasing policy of retrenchment also coincided with the official decision in December 1919 to return to the gold standard at prewar parity at some convenient future date, a decision which in itself would require relative deflation in Britain. But whether the policy reversal can be held responsible for the breaking of the boom is more debatable. It certainly killed off the social reform programme, as Morgan points out,[1] and it undoubtedly aggravated the downturn of 1920-21. Moreover, the decline in central government spending during the course of 1919 exerted a deflationary impact, though this was partly offset by higher local authority spending. However, the fundamentals were already working to undermine the boom before the new policies had time to bite properly. In particular, private consumption had begun to weaken before the peak in economic activity as consumers showed resistance to higher prices, and it is significant that consumers' expenditure declined marginally in 1919-20 compared with a 21 per cent rise in the previous year. In addition, some firms were becoming heavily indebted as a result of wild financial speculation and the banks were also getting increasingly illiquid on account of their large credit commitments to industrial and commercial enterprises. Real unit costs in industry were also rising rapidly at this time largely because of the implementation of the eight-hour day in 1919 which reduced weekly hours of work by 13 per cent, and there is little evidence that these costs pressures were fully offset by productivity improvements. Moreover, the speculative nature of the boom in the latter half of 1919 began to raise doubts in the minds of even the most optimistic businessmen, and it is therefore possible that business confidence was already beginning to wane before the government decided to reverse its policy stance. Exports too were beginning to weaken, though the turning point here followed rather than preceded the break in domestic activity.

Thus the turning point of the boom can be located in the spring and early summer of 1920. Prices peaked in the spring of that year when unemployment was at its lowest, around two per cent. Industrial activity held up a little longer partly because of the delayed response of residential construction following the implementation of the 1919 housing subsidy, but it soon responded to the squeeze on consumption and the weakening of exports in the latter half of the year. However, the speed with which the economy moved into depression, and the severity of the downturn, took most contemporaries by surprise. By the middle of 1921 Britain was experiencing one of the worst recessions in history, far more severe than 1929-32 or the early 1980s, though somewhat shorter in duration. Unemployment rose to a peak of 2.4 million in May 1921, over 22 per cent of the insured labour force (seasonally adjusted). Apart from fixed investment, which was bolstered up by the lagged effect of the subsidised housing programme, all major indices of economic activity fell sharply. Exports declined by 30 per cent, industrial production by 18.6 per cent, total output by 14 per cent and employment by 14.4 per cent. Money wages also fell sharply in 1921, partly as a result of sliding-scale agreements whereby wages were linked to price changes,[2] though real earnings actually rose modestly because of the even greater fall in commodity prices and the cost of living. The severity of the depression was undoubtedly aggravated by the government's policy of fiscal and monetary retrenchment in their effort to restore financial rectitude and pave the way for the return to the gold standard which had been formally abandoned at the end of the war. Severe labour unrest, especially in the coal and cotton industries and on the railways, added to the misery of the period.

The violent boom and slump of 1919-21 are worthy of some attention since it can be argued that many of Britain's difficulties in the 1920s and beyond stem from this episode rather than directly from the war itself.[3] As we have noted, the boom was one of prices and speculative activity rather than substantial gains in real output and productivity. In the wave of euphoria that accompanied the prospect of getting back to peacetime production manufacturers were in a good position to benefit from rising demand and the initial lag of wage costs behind prices. Unfortunately the benefits were not wisely distributed. Profit gains of war were recklessly dissipated; in particular, a misplaced faith in the future potential of the staple trades led to an orgy of speculation in assets which eventually resulted in many firms being overburdened with redundant capacity bought at inflated prices. There is little evidence that much of the investment and re-equipment of this period took place in areas of future growth potential (for example newer sectors of activity) or in modern technology. Indeed, the signs are that most of the investment went into second-hand assets and old technologies, while many of the benefits of wartime rationalisation and streamlining of production methods were quietly forgotten as manufacturers and trade unions sought to return to the antiquated practices of the past. In fact, with one or two important exceptions, little permanent benefit emerged from the wartime production experience or from the grandiose reconstruction schemes and intentions formulated in the latter part of the war. Corelli Barnett's vivid description of the situation may be a little overdrawn but it is nearer the truth than most of us would care to admit:

> When the guns ceased to fire on 11 November 1918 it was for Britain what the stroke of midnight was for Cinderella. The brilliant work of national reorganisation and development changed back to the now dried-up pumpkin of laissez-faire individualism. For what had been done in wartime was looked upon as an extraordinary response to an extraordinary situation, having no relevance to peacetime. Indeed, by a paradox, the war rendered British businessmen more self-satisfied rather than less, because, forgetting that foreign machines had alone made the colossal output of munitions possible, they took victory as being proof that British industry was superior to German.

The trade unions were no less conservative and nostalgic; if anything they were to prove an even greater obstacle to change and growth in the postwar years as they quickly resumed their former attitudes and practices. Barnett's trenchant comments are worth quoting further:

> Now that the war was over, the trade unions' consent to work modern factories in a modern way, so as to achieve the utmost efficiency and productivity was revoked. The main purpose of life, stubbornly pursued, became once again to preserve the ancient skills and prescriptive rights of their members, often by opposing the introduction of new machines and methods, but alternatively by demanding that men rendered superfluous by new equipment should nevertheless be retained on the payroll . . .
>
> The trade unions used their control of apprenticeship — in any case a

> medieval system of industrial education — not to increase the amount of skilled labour, but to restrict it. Traditional right rather than modern functions also determined those convoluted borderlines between one union and another which divided up work in a British factory like a map of Germany in the eighteenth century. The British trade union's aim was maximising manning rather than maximum efficiency and profitability and hence maximum wages. As a consequence the more go-ahead a firm or industry was, the more it was bound to run into dour, pig-headed trade union obstruction. Trade unionism was therefore particularly to blame for the fact that even new industries in Britain grew far more slowly than their foreign rivals.[4]

Looking down the decades one can scarcely fail to be struck by the frequency with which such strictures on management and labour have been used to explain Britain's economic performance. Sadly they have defied precise quantification. Returning to the more immediate scene we may take note of an area where more explicit identification can be made. If management and labour failed between them to capitalise on the benefits of their wartime experience, there was a legacy of the postwar boom which had significant longer term implications, namely the efforts made by labour to secure real income gains. The unions may not have had much time for propagating the spread of improved work practices but they certainly used their new-found strength — reflected partly in the doubling of union membership — to push up their standard of living significantly. At first there was some justification for this effort since real wages towards the end of the war were some way down on the levels prevailing in 1914. But in the next two years, by a combination of a rapid rise in money earnings and a reduction in hours of work, labour more than made good the wartime real income lag. Weekly money wages rose faster than the cost of living in 1919 and 1920, but the main improvement came through the real hourly wage level as a consequence of the general reduction in working hours in 1919 from an average of 54 a week to around 47, a decline of 13 per cent. Altogether some 7 million workers obtained reductions in hours averaging 6½-7 hours per week without loss of pay between 1919-20. The result was that hourly real rates rose strongly in 1919 and 1920, by nearly 24 per cent as against 10.6 per cent for weekly rates, and at their peak in May 1921, by which time all the reduction in hours had been completed, real hourly rates were some 41 per cent higher than in July 1914 compared with an increase of 23 per cent in weekly real rates. Because of the initial stickiness in money earnings it was not until the latter part of 1921 and 1922 that some compression of these real wage gains was achieved.

While part of the real income gain was no doubt less obvious to labour since it was taken out in the form of increasing leisure, the consequences for business were serious. Unit labour costs rose rapidly in 1919-20, as measured by the hourly rate, one reason why business became less optimistic as time went on, and they continued to do so well into the ensuing depression. They were not matched by productivity improvements and hence profits came under increasing pressure. Thus the high level of real wages in 1921 may be an important reason why the economy took time to recover, a situation which was to be repeated in the early 1930s and more recently in the 1970s and early 1980s.[5]

Secondly, the wage and price explosion of 1919-20 led to a severe widening of the price differentials between the US and UK — from 8 to 23 per cent — and this divergence was the source of the exchange rate pressure that dominated policy in the early 1920s. Whether elimination of the differential would have solved Britain's exchange problem is a moot point, but as Dowie points out: '*to the extent that price competitiveness was at the root of Britain's troubles in the 'twenties*, it can be argued that these troubles were the result of the events and policies of 1919-20.'[6] (original italics)

Dowie castigates all major interest groups for their part in the débâcle: industrialists for failing to recognise the stickiness of costs, especially labour costs, in the feverish quest for speculative profits; organised labour for extracting rewards that in a real sense were not being earned; and the government for passively validating and partly encouraging the process through the premature removal of wartime controls and the formal abandonment of the gold standard at the very time when monetary restraint was urgently requred. Whether a policy of severe monetary and fiscal retraint was politically feasible in 1919, given the social tensions and industrial unrest at the time, is a debatable point, but the fact remains that the failure to adopt such a policy until the damage was done meant that the scope for policy manoeuvre in the 1920s was thereby reduced. Moreover, the initial delay in reversing the policy stance — though the tightening of policy did begin rather sooner than some commentators suggest — inevitably entailed a rather more severe response by the government in 1920 who by then had become thoroughly alarmed by the course of events. Why retrenchment should have been any more acceptable politically and socially in 1920 rather than in the previous year seems to be something which remains unexplained, but it was clearly too harsh and too late in the light of subsequent events.

Given the sharp contraction in activity in 1921 it is not surprising that the rebound when it came was equally pronounced. However, though exports turned early, due to the depreciating pound, the initial recovery in business activity did not materialise until well into 1922. There were several reasons for this delay. Investment for one thing remained sluggish for some time partly because of the sharp fall off in housing and shipbuilding following the delayed postwar peak. Secondly, government policy remained restrictive for longer than warranted by the state of the domestic economy. It was not until April 1921 that Bank rate was lowered to 6½ per cent and even by the end of the year it was still at the relatively high level of 5 per cent. The Bank did, however, refrain from reinforcing the deflationary policy by open-market operations designed to reduce the asset base of the financial market. Even so, given the domestic situation, the Bank's policy was hardly appropriate, but then of course it was determined primarily by the weakness of the foreign exchanges. In addition, adherence to relatively high interest rates in recession reflected the Bank's desire to avoid losing control over the money market as it had done in 1919, and at the same time to ensure that the market would absorb the large volume of Treasury Bills and relieve the government as far as possible from resorting to Ways and Means Advances. Fiscal policy provided little offset either since the government was intent on restraining public spending and if possible reducing and funding the national debt. Such financial rectitude was essential if Britain was to return to the gold standard at the prewar parity as the government intended it should.

But probably an even more important reason for the delayed recovery was the continued high level of real wages, which meant adverse cost-price relationships for industry and depressed profit margins. As we have seen, real wages rose rapidly in 1919 and 1920 and continued to do so until the middle of 1921, after which they stabilised until the first quarter of 1922 and then began to fall. By the peak, real hourly rates were nearly 66 per cent above the level of July 1918 and 41 per cent higher than July 1914. Such a large increase could not be compensated for by productivity growth and/or adjustments to final product prices, at least not after the spring of 1920 when demand conditions slackened and market prices weakened. Thus through to 1922 manufacturers faced an adverse shift in their cost-price ratios which resulted in squeezed profit margins. Hence output and employment were not likely to respond until this adverse trend was reversed.

In fact the upturn in the economy corresponds quite well with the reversal of the trend in unit labour costs between 1921-22, and by the latter half of 1922 output was rising strongly. Output and industrial production recouped much of the loss of the previous year, rising by 7.4 and 15.6 per cent respectively for the year as a whole. Exports, helped by the weakness of sterling, led the recovery and rose by some 38 per cent, thereby regaining much of the loss in 1921. Consumer goods industries, especially the newer industries, also responded quickly. During the next three years the recovery continued, though at a less hectic pace, with the revival of building activity and investment carrying the process forward. Except in 1923, when coal exports were temporarily boosted following the invasion of the Ruhr by France and Belgium as a result of Germany's alleged default on reparation deliveries, there was little follow-through in exports, the volume of which remained on a plateau between 1923-25.

From the employment point of view the recovery was very disappointing. Though the number of insured unemployed was more than halved, there was still around one million out of work in June 1924, or some 9.2 per cent of the insured workforce, and thereafter the figure tended to edge upwards. Nevertheless, the strength of the recovery was quite marked between 1921-25. Industrial production rose by 41 per cent, domestic output by 18.5 per cent, exports by 51 per cent and fixed investment by just over one quarter. It should be borne in mind of course that these calculations are based on the low levels of activity recorded in the recession of 1921 and so are bound to appear impressive. Even so, by 1925 levels of activity had generally exceeded those of 1920 and 1913. Industrial output, for example, was nearly 13 per cent greater than in 1913.

In 1926 the recovery was suddenly cut short by the severe labour troubles of that year: the General Strike and the subsequent prolonged stoppage in the coal-mining industry. The downturn was much milder than that of 1921 however with declines in domestic output and industrial production of the order of 3.7 and 5.4 per cent respectively. The main burden was borne by the heavy trades, coal in particular of course, which were hit by export losses. Coal exports dropped by 60 per cent compared with a 10.3 per cent decline in total exports. Some industries continued to expand, though at a reduced rate, notably vehicle manufacturing, building, paper and printing, and gas, water and electricity.

The 1926 recession should be regarded as a temporary setback in the recovery from the slump of 1921 since it was caused by a random and exogenous shock rather than by natural forces. Had the General Strike and prolonged coal stoppage not occurred it is more than likely that economic activity would have continued on a rising trend through 1926, a conclusion confirmed by the fact that some sectors managed to make headway despite the industrial difficulties of that year. Not surprisingly, the following year saw a substantial recovery in output and exports, but thereafter the strength of the recovery tended to weaken. Compared with the United States and some European countries, Britain experienced a rather feeble boom in the later 1920s. Even the flurry of activity in the new issue market on the stock exchange was but a pale image of what was happening on Wall Street. Output increased more slowly than in the early 1920s, while in 1927-28 industrial production and investment actually declined modestly, due largely to weakness in one or two sectors such as residential building, iron and steel and shipbuilding. Unemployment remained obstinately high throughout the period at around the million mark.

The dampened nature of the boom in Britain has often been seen as a problem of exports: that export growth was checked by high and inflexible wage costs and an overvalued currency following the restoration of the prewar parity in 1925. There is some truth in the export thesis though its importance should not be overstated. The volume of exports actually rose more rapidly than industrial production between 1926-29, though much of this gain represented a recovery in coal exports following the collapse of the coal strike of 1926. However, exports continued to rise between 1927-28 when manufacturing activity was flat. Not that export growth can be regarded as entirely satisfactory taking the longer view, since export volume still remained well below the prewar level — some 19 per cent lower at the peak in 1929. On the other hand, whether the stock explanations of export failure — an overvalued currency and high wage costs — were the main reasons is another matter. A lower rate of exchange may have made things slightly easier for exporters in the latter half of the 1920s, but there is little evidence that it was the prime cause of the failure on this score (see below). But given the fixed exchange rate, wage costs assume greater significance. Money wages remained relatively stable in the late 1920s at roughly double the prewar level, and US wage costs in manufacturing followed a similar course. The big difference however is that productivity in US manufacturing was rising much more strongly than in Britain and so unit labour costs moved in favour of the former.

Several additional factors for the export lag may be suggested. International competition in manufactures increased sharply in the later 1920s as a result of the sustained recovery in European countries; in particular, Germany after the stabilisation of the currency in 1923-24, and France and Belgium, both of which were aided by undervalued rates of exchange. No doubt the delayed recovery in Europe after the war had lulled Britan's exporters into a false sense of security. Secondly, many of Britain's exports consisted of staple products — coal, ships, textiles, heavy engineering — the demand for which was either declining or rising only very slowly. These conditions meant that demand was relatively inelastic with respect to price changes and hence any lowering of the exchange rate would be unlikely to have marked effects. In any case, many of Britain's staple goods were relatively uncompetitive in world markets because of high unit costs of production arising from the technical inefficiency of the industries concerned, and also because of lack of competitiveness on non-price counts. Thirdly, the overall structure and composition of Britain's export trade and markets worked to her disadvantage. Britain specialised in exporting staple products to primary producers or low-income countries instead of high-income elasticity products to richer countries. This pattern had a twofold effect. Britain was hit badly by industrialisation in the lesser developed countries — for example, textiles in Asia — while the tendency for the incomes of primary producers to sag in the later 1920s as a result of declining commodity prices reduced the propensity of those countries to import British goods.

This unfavourable set of circumstances obviously limited the scope for export-led growth in the 1920s. But the absence of a strong boom later in the decade cannot be explained solely in terms of exports. Reference to American conditions at this point is instructive. During the 1920s the US experienced a vigorous investment and consumption boom based on construction, services, transport (cars) and the newer industries. In contrast, these primarily domestically-based industries were much less buoyant in this country. The impact exerted by building and the newer industries such as motor manufacturing, chemicals and electrical engineering, was much less marked than in the case of America. Building, for example, collapsed in 1927-28, partly because of the cutback in housing subsidies, and there was a negative growth rate in construction between 1926-29. The newer and potential growth industries, though expanding steadily, were as yet too small in total to exert a really significant impact on the economy, and in any case growth rates in some of the newer industries were no greater than the average for all industry and considerably lower than in the first half of the 1920s. Furthermore, most of the service industries, including transport and distribution, recorded only modest rates of expansion.

While the muted nature of the boom may have eased Britain's position in the ensuing great depression (see Chapter 2), the fact remains that economic performance overall in the 1920s was disappointing. It is true that prewar levels of activity had been exceeded by a reaonable margin by the end of the decade, 25 per cent in the case of industrial production and 12 per cent for domestic output, but exports were well down and compared with her major competitors Britain was slipping behind in the economic league table as she had been doing before 1914. In addition, the persistently high level of unemployment remained a permanent blot on the economic landscape. The fact was that Britain's economy was faced with a series of interrelated structural problems — declining basic industries, stagnant exports, high unemployment and regional imbalance — all of which were to get worse in the 1930s. And due to financial, external and other considerations, the policy options to deal with these problems were heavily circumscribed. It is to these issues that we now turn.

## Structural Tensions, Unemployment and Regional Imbalance

The interwar years are best remembered for their persistently high level of unemployment. The problem emerged soon after the war, following the sudden collapse of the 1919-20 boom, and by the middle of 1921 there were over two million persons out of work. The numbers fell slowly during the 1920s but throughout the decade there was rarely less than one million unemployed or some 10 per cent of the insured workforce.[7] Much worse was to follow in the early 1930s.

Unemployment on this scale and duration was something quite new. Though comparative data are not available for the prewar period, it seems likely that unemployment in the 1920s was at least twice as high as anything experienced before 1914. The regional incidence of unemployment also throws up some sharp contrasts. In broad terms the North of the country suffered very much worse than the South. Unemployment in the latter including the Midlands was about half that of the North (including Wales and Scotland), with extremes ranging from 18.8 per cent in Wales to 3.8 per cent in the South-east (1929). This in fact represented a complete reversal of the regional spread prevailing before the war. The trade-union figures, though very incomplete, suggest a rate of 8.7 per cent in London and around 4.5 per cent in the southern counties as against 2 to 3 per cent in the North, Wales and Scotland. As the figures in Table 1 demonstrate, this regional disparity between North and South was to widen during the course of the following decade.

Another striking feature of the postwar unemployment was its heavy industrial concentration. Although most trades experienced higher unemployment than in the prewar period, it was the production industries (manufacturing, mining and construction) which accounted for the bulk of it — no less than 75 per cent as against an employment share of 45 per cent in 1929. Moreover, within this sector five large staple industries, mining, mechanical engineering, shipbuilding, iron and steel, and textiles, accounted for nearly one half the insured unemployment in the middle of 1929. At a time when overall unemployment averaged less than 10 per cent, about one fifth of coalminers and iron and steel workers were unemployed, 23 per cent of shipbuilding workers and 14.5 per cent of cotton operatives.

*Table 1: Unemployment percentages by divisions (July each year)*

| Region | 1912-13 | 1929 | 1932 | 1936 | Average 1929-36 |
|---|---|---|---|---|---|
| London | 8.7 | 4.7 | 13.1 | 6.5 | 8.8 |
| South-eastern | 4.7 | 3.8 | 13.1 | 5.6 | 7.8 |
| South-western | 4.6 | 6.8 | 16.4 | 7.8 | 11.1 |
| Midlands | 3.1/2.5[1] | 9.5 | 21.6 | 9.4 | 15.2 |
| North-eastern | 2.5 | 12.6 | 30.6 | 16.6 | 22.7 |
| North-western | 2.7 | 12.7 | 26.3 | 16.2 | 21.6 |
| Scotland | 1.8 | 11.2 | 29.0 | 18.0 | 21.8 |
| Wales | 3.1 | 18.8 | 38.1 | 28.5 | 30.1 |
| Great Britain | 3.9 | 9.7 | 22.9 | 12.6 | 16.9 |
| South Britain | | 6.4 | 16.2 | 7.4 | 11.0 |
| North Britain and Wales | | 12.9 | 29.5 | 18.0 | 22.8 |

1 West and East Midlands

*Sources:* W. H. Beveridge, *Full Employment in a Free Society* (1944), p. 73, and 'An Analysis of Unemployment, I', *Economica*, 3 (1936).

The contraction of the staple industries, which was to dominate the interwar economy, was dramatic following the postwar boom and much of it can be attributed to the sudden collapse of export markets. By 1922 for example, exports of cotton textiles were less than one half the 1913 level; coal exports in 1921 slumped to one third of their peak in 1913, while shipbuilding experienced an equally sharp decline. As a consequence employment in the main staple trades fell by more than a million in the first half of the 1920s. Though there was some recovery from these low levels, none of the staple trades regained their former predominance in export markets. Their main weakness in fact lay in their heavy dependence on exports of basic products the market demand patterns for which were moving adversely as a result of the slower growth in world trade, tariffs and import substitution, increasing competition in third markets, changes in tastes and technologies and the persistence of excess capacity. Thus British cotton textiles were forced out of many lesser developed countries, e.g. India, through import substitution and competition from third parties, notably Japan. In the case of coal, world demand was barely increasing as new fuels came on stream while there was fierce competition from new suppliers. In some regions employment in mining was adversely affected by the advance of mechanisation. Shipbuilding was hampered by severe excess capacity arising from the enormous expansion of world shipbuilding capacity between 1914-20 when it is estimated to have doubled. Britain also lost market share into the bargain and so over one half of her total berths were lying idle in the 1920s.

Adverse external factors dealt a mortal blow to the staple industries but one should also recognise that they had their own internal problems which served to weaken their response and to make them neglect new opportunities. They were generally inefficient, overmanned with low productivity levels and high costs which meant a weak competititive response in world markets. Their organisation and structure also left much to be desired, with too many small units working with outdated plant and with a marked reluctance to cooperate in order to reduce capacity. A recent illuminating description of the steel industry runs as follows:

> The steel industry in the 1920s faced inescapable problems of reorganisation. A century of world dominance had been followed by fifty years of piecemeal improvements. The legacy of this, in a period of chronic depression, was an industry with too many firms and too many plants. Ownership was fragmented, equipment was a motley assembly of new and old units poorly integrated with each other and below optimum capacity. In particular, blast furnaces fell well below American and European scale and the average size of steelworks was only a fifth of what contemporary experts reckoned to be the minimum efficient scale.

> Above all iron and steel-making were rarely integrated adequately in hot metal practice with consequent diseconomies in speed, organisation and fuel economy. Where there were new modern units, particularly in steelmaking and rolling mills, low capacity utilisation pushed overhead costs up punitively. The location of the industry was based on locational choices made fifty years or more earlier to suit a very different configuration of raw material costs, notably of domestic coal and ore resources that were now nearing exhaustion. The results were high costs and collapsed profits, and, during the 1920s, an increasing flow of imported cheap continental semi-finished steel.[8]

Similarly, in the case of the cotton industry Lazonick has shown how the nineteenth-century structure inherited from competitive capitalism militated against modernisation. Horizontal competition in spinning, weaving and marketing together with vertical specialisation at all levels stifled technical re-equipment in either traditional or modern capital-intensive techniques. Many inefficient firms therefore remained in business living off their capital. Moreover, British cotton managers not only resisted structural change, as did the steelmasters, but they were in fact incapable of implementing it because their specialised skills acquired under the age of competitive capitalism were unsuited to the leadership required for the new corporate age.[9]

Similar criticisms could be made about most of the staple industries.[10] By the 1920s, if not before, they had become high cost industries whose competitive strength was unequal to the challenging conditions of that decade. What was particularly unfortunate was the fact that they had assumed such importance in the industrial structure before 1914, with the result that their sudden collapse left little to take their place. The newer sectors of development — motor manufacturing, electrical engineering, precision instruments, chemicals — had made only limited headway before the war and hence these were not in a position to fill the gap created by the contraction of the heavy trades. Furthermore, high unemployment and modest real income growth in the 1920s were not ideal conditions for stimulating the rapid expansion of these new technologies, at least not sufficient to absorb the redundant labour from the staple industries. The contrast with the United States is instructive, where rapid growth in the new sectors was assisted by higher income *per capita* levels and relatively full employment.[11]

Britain in the 1920s therefore had an enormous structural problem to contend with: the contraction of the large staple industries which had gone ex-growth and the concomitant shift of resources to the newer and/or expanding sectors of activity. It was a transition which would take time to accomplish and which appeared all the more acute given the strong regional concentration of unemployment in the northern half of the country. Five areas — South-west Scotland, the North-east coast, Lancashire, Merseyside and South Wales — accounted for nearly half the total unemployment though containing less than one third of the insured population and just over one quarter of the total population of the country.[12] It was not so much that these regions suffered from overspecialisation on a narrow band of industries — though Wales and the North of England did have relatively high coefficients of specialisation — but the fact that initially they had a high proportion of resources devoted to declining industries. In 1921 one half or more of their employment was located in declining sectors of activity compared with 35 per cent for the South-east and 45 per cent in the case of the South-west and the Midlands.[13] The contrast is even more stark when attention is confined to the localisation of the staple trades. In 1923 well over one third of the insured population of areas such as Lancashire, Northumberland and Durham, South Wales and the West Riding were engaged in the five declining industries of coal, cotton, wool, shipbuilding and iron and steel. London and the South-east, by contrast, had only about one per cent of their insured population in these industries and the Midlands 12 per cent. Conversely, London and the Home Counties and the Midlands had a fifth to a quarter of their employment in rapidly expanding trades as against 10 per cent or less in the northern areas.[14]

The disadvantages of an unfavourable industrial structure can be readily demonstrated by the experience of one or two individual regions. Wales was undoubtedly the worst sufferer in this respect. It was the only region in which the numbers in insured employment actually fell and there were no really prosperous zones which stood out from the rest of the country. Wales illustrates, all too well, the dangers of extreme specialisation on a narrow band of declining industries. In 1930 blast furnaces, iron and steel smelting and rolling, tinplate, coal and slate mining and quarrying absorbed 70 per cent of the workers enumerated in the Census of Production. The other main occupations were tourism and agriculture. The worst hit part was South Wales, especially the industrial belt, which relied on the heavy industries of coal, iron and steel and tinplate. Local dependence on one or two trades was extremely high; in 1923, 56 per cent of the insured population of the eastern sector was engaged in coal mining, while in some of the valleys such as Rhondda and Pontypridd the proportion was as high as 87 per cent. Further north the range of industries was somewhat wider though the impression of a balanced economy is rather misleading because of the strong localisation of particular industries. Adjustment to changing conditions was extremely slow in Wales compared with the other depressed areas. In fact for the most part it took the form of shifts of labour out of the contracting industries into unemployment, or at best migration to more prosperous areas. There were very few rapidly expanding industries and only one new industry, rayon in Flintshire, took root in this period. Unemployed workers had little option but to migrate if they were to stand any chance of finding work, and in fact many did move to the newer industries in the South, especially to motor manufacturing in Oxford and Slough. As a consequence Wales was the only region to experience a net loss of population through migration.

Though Scotland suffered severe problems in this period it was not quite as badly hit as Wales, since it had a relatively prosperous sector in the east around Edinburgh, Perth, Aberdeen and parts of the Lothians. It was the main industrial belt in the South-west and the Clyde region that bore the brunt of the depression. Again the problem was one of the high proportion of the region's output which came from the vulnerable, export-orientated staple trades. Just over 40 per cent of all workers were employed in mechanical engineering, shipbuilding, coal, and iron and steel, as against a national average of 25 per cent, while Scotland's share of the new growth industries (vehicles, electrical engineering, rayon, non-ferrous metals, paper, printing and publishing) was only 8.3 per cent compared with an average of 14.1 per cent for the whole country (1924). Scotland's difficulties stemmed primarily from the dramatic collapse of export demand for her once staple products. The volume of exports through Scottish ports fell by no less than 56 per cent between 1913 and 1933.[15] Heavy concentration on declining basic sectors resulted in lower productivity and lower income *per capita* than the rest of the country and this in turn discouraged the development of new and expanding lines of development.

The North-east coast, including Northumberland, Durham and the Cleveland district of Yorkshire, had the distinction of having the heaviest concentration of unemployment in England. Again the main cause was the overcommitment to industries either in decline or extremely sensitive to the trade cycle. In 1924 a group of six heavy trades — coal mining, shipbuilding and marine engineering, iron and steel, heavy chemicals, shipping and port services and general and constructional engineering — accounted for over 64 per cent of all insured workers. The most severe declines in employment took place in coal and shipbuilding. In the case of the latter it was mainly due to the rapid fall in output associated with the contraction of the export market; the fall in coal output, on the other hand, was more modest and here the chief factor making for reduced employment was the rapid mechanisation of mining operations, which raised efficiency.

Finally, Lancashire is worth considering because of its close association with cotton textiles. The county was not, however, uniformly depressed since in the south around Manchester there were parts that were equally as prosperous as the South-east. But conditions in much of the north of the county were almost as bad as those in Wales and Scotland because of the heavy reliance on cotton, coal and general engineering. This group of industries accounted for about 35 per cent of all insured workers in 1923. The areas worst affected were the Barrow district, dependent on shipbuilding and iron and steel, parts of Merseyside including Wigan, Warrington and St Helens (coal) and above all, the chief centres of the cotton trade — towns such as Blackburn, Burnley, Rochdale, Oldham, Bolton, Preston and the Rossendale Valley. The overwhelming importance of the cotton industry to these towns resulted in unemployment levels almost as high as those experienced in South Wales. In Blackburn, for example, cotton accounted for 60 per cent of employment, and the unemployment rate for the town in 1931 was 46.8 per cent.

The industrial structure of the Midlands and South, the latter especially, contrasts markedly with that of the depressed northern regions. The most notable features of the South's industrial structure were its marked diversity and lack of declining heavy industries. London of course was the classic case of a well-balanced diversified structure. Nearly every main industrial group was represented in the metropolis, the highest concentration of employment being in building and hotels and clubs (8 and 6.2 per cent respectively of the insured population), both of which were expanding rapidly. London was not only an important manufacturing centre with a wide range of industries, including electrical engineering, food products, light metal trades, vehicles, scientific instruments, chemicals, leather, paper, furniture and clothing, none of which were unduly weighted in the industrial structure, but she also had the additional advantage of being the centre of Britain's financial, commercial and administrative services. It was this great diversity coupled with the absence of heavy basic industries and the high proportion of employment in non-manufacturing activities that was largely responsible for London's prosperity. As Fogarty commented, 'London is ... an outstanding example of the advantages to employers and workers alike of the concentration in a single area of a wide range of growing and adaptable manufacturing industries backed up with highly specialised and efficient technical, commercial and financial facilities.'[16] Likewise, many of the south-eastern counties shared in this prosperity and one can again point to their favourable industrial structure as an important beneficial factor.

Thus much of the difference in prosperity between the North and South of the country stemmed from the initial pattern of the industrial structure. The North had far too many resources locked up in the basic industries and other declining sectors where the growth potential was exhausted, whereas the South was favoured with a diversified structure containing a predominance of growth industries and very little in the way of heavy industry. Thus to make good their initial deficiency the growth of the expanding industries in the less favoured regions would have had to have been very much faster than the national average, and/or they would have had to attract a larger share of the growth industries. Neither in fact happened. The rate of increase of the expanding sectors was no faster than elsewhere in Britain, while the rate at which new activities were attracted to these regions left much to be desired. Consequently, they were overwhelmed by the forces of structural imbalance.

In fact, because of their structural imbalance and other related factors, the cards were stacked against the northern regions. They could not therefore generate above average rates of expansion in the non-basic sectors, nor did they offer a particularly attractive location for new developments. The uncongenial environment, lower incomes and market forces conspired against them. It was especially unfortunate that these areas faced a collapse in demand for their traditional products at the very time when the market was becoming a more powerful determinant of industrial location. During the nineteenth century the North had had two great assets: good rail and sea communications and the availability of key raw materials, especially coal and iron. But the locational strength of these two factors was considerably weakened with the development of motor transport and electricity which made industrial location much more flexible. Access to markets therefore became the main influence determining the location of many of the newer and expanding trades, and since the South had the twin advantages of a large and growing population and a higher level of purchasing power it was only natural that the growth industries should gravitate to this part of the country. The process became a cumulative one: new industries and firms were established, incomes and employment grew, population expanded through migration and natural increase, and in turn the market expanded, which stimulated further development. In the North the reverse sequence operated, added to which the less pleasant environmental conditions, the more militant and traditional workforce and the dominance of one or two firms or industries, tended to deter the rate of in-migration of new firms. Moreover, the interdependence of many new trades in terms of the demand for and supply of their products meant that there were obvious economies to be realised from geographical concentration at the initial locations. Thus, for example, firms producing components for the motor industry tended to congregate close to the assembly plants in Oxford and Luton rather than in the North since this reduced transport costs to a minimum. In short, given the unfavourably weighted industrial structure at the start of the period the depressed regions found it difficult to diversify and adjust their economies, especially as they faced competition from the more firmly established growth industries in the South.

Thus much of the heavy unemployment which emerged in the 1920s and which was to get worse in the following decade can be seen as a long-term persistent problem arising from the severe structural tensions in the economy. The problem appeared even worse than it actually was because of the strong industrial and geographic incidence of unemployment which exposed the vulnerability of the northern regions with their excessive specialisation and dependence on a narrow range of declining sectors and their limited base of potential growth industries. External forces were largely responsible for the sudden transformation of the basic sectors, though these industries were not helped any the more by their own inefficient and high-cost operations arising from a combination of factors including low productivity, poor organisation, antiquated technology and high real labour costs. There was no way of preventing the process of structural change and the attendant unemployment, short of directly subsidising employment in the sectors affected, since no amount of policy action could stem the advance of market forces. Adaptation might have been eased by labour transfer schemes and locational policies but neither could transform the situation in the short term. In the event, there was little in the way of regional policy direction except for a belated attempt at assisting labour migration, in the first instance from the coalfields, while macro-policy was determined largely by considerations other than those of unemployment and structural transformation, as we shall see.

## The Policy Dilemma in the 1920s

Given a situation in which factors of production are severely underutilised, a postwar (post-1945) policy prescription would be one of reflation. In the 1920s this did not materialise either on the fiscal or the monetary front. It is important to determine why it did not do so because the conventional explanation — that policy action was constrained by the pursuit of financial rectitude *per se* by hard-headed Treasury and Bank officials steeped in pre-Keynesian economics — is not really very informative unless one understands the economic background to the actual policy approach in this period.[17] It is true of course that the conventional wisdom handed down by classical and neo-classical economics together with the absence at that point in time of a clearly formulated and viable alternative economic analysis, obviously considerably narrowed the range of view of the policymakers. Yet that alone did not preclude an easier or different policy stance had conditions and priorities permitted such. After all, the 1930s were to bear witness to a rather easier credit policy, while the use of government-financed public works for employment purposes already had a reputable lineage going back to at least Elizabethan times, and such works were to be employed sporadically in the interwar years. The first world war had also given policymakers plenty of experience with deficit financing, though one could argue of course that the upshot of this experiment acted as a constraint to radical action at a later date.

Rather, the simple fact was that there were considered to be more urgent priorities than that of employment creation for determining the direction of policy. The first of these was the question of currency stabilisation: first the quest to get back to the prewar sterling parity, and subsequently the struggle to maintain it. This necessitated a hard monetary policy and, by implication, precluded an active fiscal policy. But in the latter context there was a second impediment to radical action, namely the problem of debt management following the explosion in the national debt during the war. This in itself restricted the scope for an expansionary fiscal policy. It should be borne in mind that neither the Bank of England nor the Treasury were totally unmindful of the needs of the domestic economy, a fact which they demonstrated on more than one occasion. But since it was their belief that a satisfactory solution to the currency and debt problems was an essential prerequisite to any lasting and sound recovery, these objectives took precedence in the formation of economic policy during the 1920s. Financial stability, in the broadest possible sense, was deemed to be essential if the real economy was to thrive, a view reinforced by the disastrous consequences of the inflationary explosions in several European countries in the early 1920s.

Throughout the 1920s therefore, the Treasury had a convenient excuse for avoiding radical experiments in fiscal policy, namely the urgent task of dealing with the huge national debt and the concomitant need to restrain public spending. Its size and servicing costs dictated a policy of strict financial retrenchment. The method of financing the war — principally through borrowing rather than taxation — had led to a massive rise in the national debt. The combined deficits on the war budgets amounted to around £7,186 million, and the total national debt increased by a similar amount to over £7.7 billion as against a mere £0.6 billion in prewar years. The annual servicing costs were particularly heavy since much of the borrowing had been contracted at relatively high rates of interest. The war bond of 1917, for example, had been issued with a coupon of 5 per cent, while the average interest payable on the total war debt stood at 4.65 per cent. As a consequence, debt service charges rose from 11 per cent of central government spending in 1913 to 24 per cent in 1920, and to more than 40 per cent by the end of the decade, while as a proportion of GNP they rose from one per cent prewar to 7 per cent in the 1920s.[18] An additional problem was that a large slab of the debt — some £1.5 billion by mid-1919 — consisted of short-term floating debt, which posed difficulties in the money markets in the early postwar years.

Not surprisingly the enormous debt burden proved a constant source of anxiety to Treasury officials in the 1920s, which explains why their principal goal throughout the period should be that of reducing the overall costs of servicing the debt by conversion and funding at lower rates of interest. A second task was to ensure that the overall debt did not rise, or better still to arrange fiscal accounting in such a way as to allow some of it to be paid off. Unfortunately, only very limited headway was made with the first objective partly because external considerations dictated a level of interest rates somewhat higher than suited the Treasury's purpose. Thus for much of the time the Treasury could be found censoring attempts by the Bank of England to raise interest rates on the grounds that this would jeopardise the funding programme. Though some funding of the floating debt was secured after 1921 when interest rates fell, this only amounted to a small proportion of the total debt (£1,000 million). Thereafter the market remained unfavourable for low coupon funding and conversion. Hence by the later 1920s the debt problem loomed as large as ever; most of it remained contracted at high rates of interest, there was still a substantial floating debt, and even more to the point, the cost of servicing it was higher in real terms and as a proportion of central government spending than it had been previously. It was not until the big conversion operation of the early 1930s that the debt was put on a more secure basis and its servicing costs reduced.

Having been thwarted in its attempt to reduce the size of the debt bill, the Treasury turned its attention to an area where it could expect to exert greater direct control, namely its own budgetary operations. If it could not reduce substantially the servicing costs by funding or conversion at least it could make sure that budgetary policy did not lead to any increase in the size of the debt through deficit financing, and with luck if budgetary surpluses could be secured then some of the existing debt could be repaid. Hence the Treasury's second objective came increasingly into play during the 1920s, that is to maintain rigid control over public expenditure. In this respect the Treasury achieved greater success than on the debt management front.

Initially however there was a far more urgent reason for regaining control over public spending. Government spending had been running at a very high level throughout the war, with large budgetary deficits, and these conditions carried over into the immediate postwar period with highly inflationary consequences. Public outlays accounted for about one half of total national expenditure and the budgetary deficit for the fiscal year ending April 1919 was no less than £1,690 million. This lax fiscal stance undoubtedly helped to fuel the inflationary boom of 1919 and prompted a curtailment of state spending in the latter half of that year with the result that by the fourth quarter of 1919 the budget was running in surplus, though for the fiscal year as a whole a moderate deficit was recorded. In the following year (1920-21) retrenchment was carried further: spending was cut, taxes raised and the outturn for the fiscal year was a modest surplus. The timing of the budgetary clampdown was particularly unfortunate given the marked deterioration in economic activity after the spring of 1920. Nor was there any let-up on the fiscal front during the next year or two. Deflationary budgetary policies were pursued in an effort to achieve a surplus to pay off the national debt and also to help pave the way for the return to the gold standard. Though some taxes were reduced — excess profits duty was abolished and the standard rate of income tax was lowered from 6s to 4s 6d in the £ between 1921-23 — these reliefs were offset by sharp cuts in most categories of expenditure, especially on social services, following the famous Geddes Review of 1922. Thus through to 1923 central government spending fell by nearly one quarter from the 1920 level. The position might have been a good deal worse had local authority spending not risen steadily in 1921 and 1922 partly through what turned out to be a fortuitous delay in the start of the housing scheme under Addison's Act of 1919. Consequently, total public spending (central and local) fell only slightly in real terms in these years, and as a proportion of GNP it was higher than in 1920. In other words, the high level of local authority spending helped to take some of the sting out of the government's deflationary policy, at least until 1923.

In 1924 the new Labour Chancellor attempted to reverse the severe deflationary thrust by reductions in taxation, and this policy was continued by his successor Churchill. Direct taxes were reduced, industrial firms secured rate relief in 1929, while there was a substantial increase in social welfare expenditure. Most of these items were covered by increased tax yields, by new imposts, for example excise duties and a tax on betting, and by further economies in expenditure. And with the help of Churchill's 'window-dressing' devices the budgets for the later 1920s remained more or less in balance.[19] Expenditure of both central and local governments rose in the late 1920s, though remaining fairly stable as a proportion of GNP, while the combined budgets of all public authorities were in modest surplus except during 1926. On balance there was a slight redistribution of income in favour of the lower classes.

Overall therefore the Treasury was fairly successful in keeping a tight rein on the budgetary accounts in the 1920s. This meant that it was at least possible to contain the size of the national debt though it did not leave much leeway for reducing it. On average it produced a sum of about £50 million a year for purposes of debt reduction, a very small amount in relation to the total sum. Thus despite its efforts the Treasury had not been able to make much impression on the size of the debt by the end of the decade.

On pragmatic grounds therefore, the Treasury had a strong line of defence for its policy of fiscal rigour. At times it may have gone to excessive lengths to secure economies, particularly in the early 1920s when unemployment was very high, but over the longer term it did not have a great deal of room for manoeuvre. The vast scale of the postwar debt left little scope for a large programme of fiscal spending, since this would have pushed up debt levels and servicing costs further with consequent adverse effects on interest rates and financial markets. Unbalanced budgets and deficit financing would have undermined confidence at home and abroad which was the last thing the government wanted to happen given its commitment to getting sterling back to parity. Moreover, an interesting income distributional point is often overlooked. Insofar as the bulk of the debt contracted at high interest rates was held by the higher income groups, there was obviously an equity advantage in reducing the debt and/or funding at lower rates, thereby reducing the income accruing to wealthy debt holders. A rise in the debt through resort to deficit financing would have brought benefit to wealthy debt holders.

Curiously enough the Treasury sought to defend its stance on public spending not on the practical grounds of debt management, but in terms of the inability of higher government spending to alleviate depression and unemployment. While this attitude owed more than a little to pre-Keynesian economic thought, it became something of a dogma in the 1920s, known as the 'Treasury View'. In his budget speech of April 1929 Churchill provided a summary of the prevailing ethos: 'It is orthodox Treasury dogma, steadfastly held, that whatever might be the political and social advantages, very little additional employment can, in fact, and as a general rule, be created by State borrowing and expenditure.'[20] There were several different strands of thought enshrined in this view. The most popular conception was that public spending would simply crowd out private expenditure, under the assumption that involuntary unemployment was at best a temporary phenomenon and in the long term did not exist. Secondly, it was argued that public works employment which did not yield a proper rate of return led to a wasteful misallocation of resources. Thirdly, public works could lead to distortions in employment through industrial and regional mismatches in the labour market. Broadly speaking therefore, the wider issues of fiscal policy were not considered relevant to the unemployment problem. Whether in fact fiscal policy could have provided a complete and viable solution to the problem, given its peculiar structural characteristics and bearing in mind the economic constraints, is debatable. It is an issue to which we shall return in due course.

If the budgetary issue became very much an accounting exercise by virtue of the constraints imposed by the debt problem, monetary policy was equally constrained by external considerations. Throughout the 1920s the currency question was accorded top priority: first the struggle to get back to the prewar parity and subsequently the effort required to maintain it until the abandonment of the gold standard in September 1931. Given the weaker state of both sterling and the balance of payments after the war, these objectives required a tighter monetary policy than would otherwise have been the case. In this sense then, external factors conflicted with domestic needs in a way that had not generally been the case before 1914. Thus apart from the impact of monetary policy on the domestic economy, one must also consider whether return to the former parity in 1925 put Britain at a competitive disadvantage.

The currency issue of the 1920s really begins with the first interim report of the Cunliffe Committee (on Currency and Foreign Exchanges after the War) published in August 1918. The committee's main task had been to consider how and when Britain should return to the gold standard. The main recommendation was unequivocal: that Britain should return to the prewar parity at the earliest opportunity. No other alternatives were contemplated (for example devaluation) since it was widely felt that the stability of not only Britain but the entire world economy was dependent on sterling's return to par. The committee did recognise that there would be adjustment problems on account of the deterioration in Britain's competitiveness, but it was anticipated that these could be solved by adjustment in relative price levels. To facilitate matters on the home front, it advocated cutbacks in the floating debt and budget deficits, a restriction of the note issue and a reduction in Ways and Means Advances.

Monetary policy soon came to be dominated by considerations of restoring the sterling standard to its original value of $4.86. Initially however the question of the return to gold was not the key determinant of monetary policy. Although restoration of the standard was accepted as official policy in December 1919, there was no question at that stage of an early return. When the artificial wartime support for the currency was withdrawn in March 1919 (the formal abandonment of the gold standard), the pound fell sharply on the exchanges to $4.02 in November 1919 and eventually to a low of $3.20 in February 1920, as against the wartime pegged level of $4.76. Given the size and speed of the fall it would have required an early and very severe restriction of credit and high interest rates to have pulled the pound back to anywhere near its former level, and the authorities were not at first prepared to countenance such premature action during the period of demobilisation and reconstruction because of the social and political implications. However, as we have seen, the authorities were forced to call a halt to the monetary accommodation policy earlier than they really wished and before the reconstruction phase was complete, since the boom of 1919 appeared to be getting out of control. Two features caused particular concern: the high inflationary potential running at about 20 per cent per annum, and the Bank's loss of control over the money market. Consequently during the latter half of 1919 the authorities steadily reversed their previously accommodating policy stance. Government spending was trimmed, a decision was taken to limit the Treasury note issue, while in November 1919 Bank rate was raised to 6 per cent (it had stood at 5 per cent since April 1917) and subsequently to a peak of 7 per cent in April 1920. Once having started on a deflationary course it became a convenient excuse to continue it for some time, not only to clear the market of its inflationary potential, but also as a platform for preparing the way for the return to gold. Interestingly enough, the breaking of the boom by means of dear money was advocated by no less an authority than Keynes, who suggested that interest rates should be kept at a high level for a considerable period of time.[21]

As it turned out this was not one of Keynes's most judicious pronouncements. Despite the severe recession which followed, the authorities maintained a restrictive monetary policy for some time partly to ensure that inflation was squeezed out of the system and partly to eliminate the turmoil in the money market. Not until April 1921 was the discount rate reduced to 6.5 per cent and it was to be another year before it reached a low of 3 per cent. During that time (1920-22) the money stock declined by 2.6 per cent and UK wholesale prices fell faster than those of the US. One favourable effect of this severe deflation was that the sterling-dollar exchange rate improved, reaching a peak of $4.72 in February 1923.

It was about this time, having regained control over the money market, that the Bank began to give serious consideration to the conditions required for an orderly return to the pre-1914 parity. The strengthening of the exchanges in the winter of 1922-23 offered the promise of an early return but these hopes were soon dashed by a renewed weakening in sterling during the course of 1923 which brought the rate down to $4.36 by December. Despite a continued squeeze on the money stock and a further fall in nominal wages during the year, the sterling exchange was adversely affected by both the failure of UK wholesale prices to improve further relative to those of the US, and an interest rate differential in favour of New York. It became clear therefore that if the pound was to be returned to its former level it would need to be forced up artificially. Consequently Bank rate was raised to 4 per cent in July 1923 and maintained throughout 1924 when US interest rates fell. The interest-rate differential therefore shifted in favour of London and helped to strengthen the pound. The return of a Conservative Government in October was a further beneficial influence, while speculation on the prospects of an imminent return also assisted the appreciation. The time seemed ripe to take action and in the following March Bank rate was raised to 5 per cent to ensure a sufficient flow of funds in from abroad to support the parity move which Churchill announced in his budget speech a month later.

Once secured, the task of maintaining the gold standard was no less problematic for the monetary authorities than engineering its return had been. The overvaluation of sterling in relation to other major currencies and the inherent weakness of Britain's external account — the current account surplus was insufficient to match the deficit on the capital account (long-term lending abroad) necessitating dependence on short-term capital inflows — meant that the exchange rate was under constant pressure and the pound remained under par for much of the time. This meant that the Bank faced periodic pressures on its reserves, notably in 1927, 1929 and 1931, the last occasion proving fatal. However, until the later 1920s the Bank managed to avoid putting undue strain on the domestic economy. Initially the position of sterling was eased somewhat by an outflow of capital from the US and by close cooperation between the Bank of England and the New York Federal Reserve Bank. As a result Bank rate remained fairly stable at around 4.5 per cent until the beginning of 1929. As far as possible the Bank tried to avoid disturbances to domestic conditions by open-market operations, thus ensuring that the cash base of the commercial banks remained reasonably stable. In addition, the Bank sought to protect industry from short-term interest-rate variations by various technical devices to strengthen its reserve position, including the prewar practice of influencing gold movements by small variations in the price of gold as an alternative to discount rate changes and by maintaining a narrow differential between market rates and Bank rate.

Unfortunately the Bank's accommodating stance had to be abandoned later in 1928 and 1929 when things became more difficult. The Wall Street boom led to a rise in the Federal Reserve discount rate and interest rates generally tended to harden. This put severe pressure on the sterling exchange and resulted in a sharp loss of reserves, providing a foretaste of what was to come in 1931. By February 1929 the Bank's gold stock, which reached a peak of £173.1 million in September 1928, had been reduced to £150 million. Bank rate was promptly raised to 5.5 per cent but this failed to check the drain. By September the gold reserves were down to £131 million, the lowest level since the return to gold, and a further one per cent rise in Bank rate had to be made at the end of the month. The domestic situation certainly did not warrant such a high rate but there was little else the Bank could do short of abandoning the gold standard. The immediate cause of the crisis was withdrawal of short-term funds, on which Britain depended to balance her accounts, as a result of the attraction of higher interest rates abroad, and there was little prospect of these being stopped unless action was taken. The high rate of interest eventually stalled the outflow and by the end of the year the gold reserves had been restored to £150 million. A progressive relaxation of Bank rate then became possible as interest rates in New York declined with the onset of depression.

## Policy Assessment

Given the economic objectives of the authorities in the 1920s, the macroeconomic policies pursued can be judged to be logically sound. The problem was that they conflicted with the needs of the domestic economy, where resources were badly underutilised. The most glaring manifestation of the domestic problem was the persistently high level of unemployment concentrated in the export-based staple trades with a strong regional bias. The main issue we therefore need to examine is in what way the economic policies of the period influenced the outturn of the real economy, and whether alternative policies would have improved matters significantly.

It is appropriate to begin first with the return to gold since the difficulties involved in maintaining the parity between 1925 and 1931 severely limited the range of policy options for dealing with unemployment. A large monetary and fiscal expansion would not have been consistent with the fixed parity because it would have undermined the balance of payments. Moreover, the sterling rate has frequently been criticised on the grounds that it was overvalued and thereby made Britain's exports less competitive.

The debate on the effects of overvaluation has been a lengthy one and not very conclusive. Many writers would agree that it was not beneficial to the economy, but one needs to be careful about stressing unduly its adverse effects. It is generally accepted — after Keynes — that the pound was overvalued by about 10 per cent and on this basis various estimates have been made as to the consequences of this.[22] One of the most pessimistic is that by Moggridge. On the assumption that the sterling price elasticity of demand for imports was -0.5 and the foreign price elasticity of demand for British exports was -1.5, he proceeds to calculate the impact on British trade and the balance of payments of an implicit appreciation of sterling of 11 per cent between 1924-25 (that is from $4.40 to $4.86 to the £) and a hypothetical devaluation of 10 per cent in 1928 (from $4.86 to $4.38 to the £). In the former case he finds that the balance of payments on current account would have deteriorated by roughly £80 million, while in the second instance it would have improved by as much as £70 million. In both cases the bulk of the change is estimated to flow from the trade balance, and in particular from the side of exports which would have declined by 10.5 and 14.1 per cent in volume and value respectively in 1924-25, and risen by 9.0 and 13.3 per cent in the later 1920s. The relief afforded by a 10 per cent devaluation would then have been sufficient to allow employment to be hoisted by 729,000, leaving a residual of over half a million or 4.7 per cent of the workforce in 1928. While the author recognises that a lower exchange would not have solved all the problems of the British economy, it would, he feels, 'have provided a better basis on which to solve these problems which centred around the transition of the industrial structure from the nineteenth to the twentieth century.'[23]

These conclusions require qualification on several counts. For one thing the elasticities used for exports and imports are probably too high. Alternative calculations by Dimsdale, using more plausible elasticity values, suggest an employment effect of 450,000.[24] It is certainly difficult to believe, given the structure of the export trade, that the trade balance would have improved to the extent envisaged by Moggridge had the pound been 10 per cent lower. In fact the return to parity only raised export unit values by 5.8 per cent, and by 1928 most of the effects had worn off. It is true that exports still remained uncompetitive but this was due more to the unfavourable cost-price structure in British industry and slack trading methods than to the rate of exchange. It is highly unlikely therefore that a 10 per cent lower exchange rate would have done much to boost the exports of the staple industries, which were after all still the major exporters, since these were checked by a series of adverse factors including unfavourable long-term shifts in demand patterns and domestic inefficiency. As Alford has noted, many British goods were not wanted at any price in international markets.[25] The newer industries would no doubt have derived some benefit from a lower pound but as yet their contribution to total exports was still very small.

Moreover, if the exchange rate was the crucial factor in Britain's weak performance in the later 1920s, one would have to inquire why it was that the export record was not better before 1925 when, under a floating exchange, the pound was on average 10 per cent or more below its former dollar parity. There is no evidence that Britain's trading performance improved markedly during the periods when the currency was most depreciated. Secondly, Britain's trading difficulties in the later 1920s were compounded by the undervaluation of several currencies, notably the Belgian and French. Any attempt by Britain to counteract the handicap by lowering the rate of exchange or by returning at a lower rate in 1925 would probably have inspired competitive depreciation moves on the part of these countries. A more telling point has been made by Drummond. Since a high proportion of Britain's exports went to Imperial countries whose locally issued currencies were tied to sterling, Britain would not have improved her competitive position *vis-à-vis* countries such as India or Australia by making the pound cheaper against the dollar or mark.[26]

The overvaluation of sterling in 1925 cannot therefore be regarded as a major cause of Britain's economic difficulties in the later 1920s. These originated well before the parity was restored and were largely the product of depressed demand for the products of her staple trades. The restoration of the sterling parity did not of course help matters but it is inconceivable that a rate 10 per cent below the former par level would have restored their health or led to a marked recovery in output and employment. As Drummond has observed: 'Monetary and exchange policies were certainly inconvenient, but no conceivable policy would have obviated all Britain's difficulties in the 1920s, because these stemmed so largely from the nation's own industrial past and from overseas developments that Britain could neither prevent nor control.'[27]

It is perfectly possible of course for an overvalued exchange rate to depress activity below the full employment level, but for the circumstances of the 1920s it is very difficult to conceive of a realistic exchange rate. Given the size and the unusual structural spread of unemployment it would have required a very much lower exchange rate than 10 per cent below the adopted parity, which is generally assumed to be the degree of overvaluation, to have made much impact. Any such move to an appreciably lower rate would have raised cost and inflationary implictions and certainly it would have invited retaliation from other countries. In any case, it is doubtful even then whether it would have brought marked improvement to the staple industries, and if it had done then one could argue that it would have retarded the vital process of structural adjustment and capacity destruction in the ex-growth staple industries. Clearly there would have been trade benefits for the new and growing sector of the economy but in the short-term at least the effect would have been too small to absorb the redundant labour of the old industries, and there would still have been a severe locational problem in terms of the demand for and supply of jobs which no amount of tinkering with the exchange could have cured.

One might therefore argue that the alleged 10 per cent overvaluation of the exchange was neither here nor there as far as the structural adjustment problems of the economy were concerned. The fact is that the viability of a substantial part of Britain's economy had been undermined by adverse shifts in patterns of world demand after the war, and exacerbated by an internal cost structure which left many industries uncompetitive. While the sterling parity did not help matters, it is implausible to suggest that a more realistic exchange rate would have obviated these twin problems.[28]

The plain fact is that Britain fared very badly in international trade during the 1920s even though world trade was expanding quite rapidly. Apart from short-term fluctuations, exports remained stagnant throughout the decade and by 1929 they were still 19 per cent down on the 1913 level. The volume of world trade, on the other hand, rose by no less than 27 per cent over this period. Inevitably, therefore, Britain suffered trade share loss; her share of total world exports declined from 13.9 to 10.8 per cent through the years 1913-29, and her share of world manufactured exports from 29.9 to 23.6 per cent.

In many respects Britain was more vulnerable to shifting patterns of world demand because of the specific structure and composition of her export trade. A large proportion of Britain's exports in 1913 consisted of textiles, coal, basic engineering products, railway materials and ships, all of which were declining sectors in world trade during the 1920s. The industrialisation of new countries and rising tariff barriers led to import substitution on the part of former customers, while the demand for some products, e.g. coal, was flat because of the development of substitutes. Furthermore, the concentration of exports on low-income markets tended to dampen the growth of Britain's exports. Some two thirds of all her exports went to primary producing or semi-industrial countries, markets which were poor from both an income and import substitution point of view. The incomes of some primary producing countries were depressed in the later 1920s because of oversupply problems, while Britain was easily the main loser from the process of import substitution in the semi- or newly industrialising countries. Her losses in these markets were substantially greater than those of Western Europe and the United States, while the major losses through import substitution occurred in India, Latin America and Japan with the main product affected being textiles.

While Britain was no doubt more vulnerable to trade losses than other countries because of the commodity composition of her exports and the markets served, these factors cannot explain fully Britain's weakness as an exporter in the 1920s. Overcommitment to staple products and traditional markets obviously proved a serious handicap, but there were general market weaknesses to be considered. Evidence suggests that there was a significant deterioration in British industry's ability to compete in world markets across the board. If this had not been the case then it would be difficult to explain the following facts. Why, for example, Britain made so few trade gains in the rich and expanding markets of North America and Western Europe during this period. Exports to North America rose only slightly while those to Europe fell by 20 per cent over the years 1913-29. Such losses cannot be explained by tariffs and import substitution, since world trade was then expanding and industrial countries were increasing their trade with each other. Or again, why in some of her formerly strong markets Britain was being pushed out by alternative suppliers. For example, in 1913 over one half of Japan's imports were British whereas by 1929 the share had dropped to one quarter, suggesting a significant switch to alternative suppliers, in this case the United States. And in the case of the three relatively rich Dominion countries of Australia, New Zealand and South Africa, Britain's share of imports fell from three quarters to three fifths between 1913-29.

If Britain had been competitively strong it is unlikely that her trade share losses would have been so heavy or so widespread. Share losses in some sectors were only to be expected, especially in staple products, which were affected severely by import substitution in developing countries and/or stagnant demand. But the fact remains that in nearly every industrial category, whether expanding, declining or stable from the point of view of world trade, Britain's share declined between 1913-29. Britain's share of world trade in 12 out of 16 major commodity groups contracted; the losses in some cases were substantial, and even more significant they occurred in all expanding groups. Price and non-quantitative information support the view that widespread trade share losses were a consequence of a general deterioration in competitiveness. An index of export unit values (all manufactured exports) shows a value of 159 for the UK in 1929 (1913 = 100) as against a weighted average of 134 for 12 major industrial countries. British export prices were higher relative to the prewar base than those of any other country and the price disadvantage affected all main commodity groups, though it was especially pronounced in machinery, transport equipment, textiles, clothing and other manufactures.[29] British trade methods also came in for criticism during this period but since this was a perennial complaint it is difficult to know how much importance to attach to it. It was certainly an additional competitive handicap and given the increase in world competition and the greater sophistication of products it may have been of more importance than in the past.[30]

Cost pressures were therefore an important factor in the decline of Britain's competitiveness, both in the staple and the growth industries. Since British industry faced higher relative unit costs prior to the adoption of the prewar parity the latter clearly cannot be the sole reason for the handicap. The fact was that Britain's internal costs were too high, with or without the sterling parity, and the only possible solution to this dilemma was either a marked improvement in the level of productivity or a compression of wage costs. The former is difficult to achieve in the short term, though some improvements took place through labour shedding in the early 1920s. The latter ran up against the increasing rigidity of wage costs in a downward direction as unions were able to resist trading wages against employment, even in the depressed industries and regions, so that there was little tendency for industrial or regional wage variations to widen. The strength of organised labour and the level of unemployment benefit made it difficult to achieve either a general wage compression or a greater regional spread of wages in accordance with market forces.

When we turn to the wider issue of macro-policy, the situation is more complex than it appears at first sight. Glynn and Booth have usefully crystallised the issue in their 'dual economy' or two sector approach to the British economy, which has significant implications for the type of policy response. 'The traditional British industrial economy, based on the staple export industries, faced a severe cost problem resulting in classical unemployment which could only be solved unilaterally by deflation and wage reductions; whereas the potential growth industries required reflation and rising consumer demand to generate expansion and development.'[31] The fact that the industrial-structural problem was translated into a regional problem, with Outer Britain (the North) performing badly in all industries relative to prosperous Inner Britain (the South), tended to complicate the policy dimension since any move towards expansionary policies to assist the latter would not necessarily help the former in overcoming its problems.

As things turned out in the later 1920s, macroeconomic policy did not help either very much. There was no general policy of expansion, nor any marked deflationary exercise to force a compression of the cost structure in the old industrial sector. Nor was there much in the way of regional measures of assistance except for the start of the industrial transference scheme in 1928 to encourage labour mobility and *ad hoc* measures to assist the old staple industries. Fiscal policy remained broadly neutral in the later 1920s following the big contraction in public spending in the early 1920s, while monetary policy was midly expansionary. There was no sharp reduction in the money stock as there had been between 1921-25 when it fell by nearly 10 per cent; in fact it rose faster than nominal GDP (4.6 as against 1.7 per cent, 1925-29) while bank deposits expanded accordingly. Internal costs adjustments were certainly in evidence in the first half of the decade when deflationary forces were powerful, but in the latter part of the 1920s this was no longer the case. As Dimsdale observes: 'The monetary policy followed by the Bank of England after the return to gold was not sufficiently vigorous to bring about the internal adjustment required to correct overvaluation. Had such an adjustment been attempted by more severe deflation, it is doubtful whether it would have been successful, because of the limited flexibility of money wages.'[32]

Monetary policy of the period has been criticised more on the grounds of its adverse effects on credit and interest-rate structure rather than for its cost compression properties. In particular, the historically high interest rates and the uncertainty created by rate changes are held to have discouraged investment. The Macmillan Committee on Finance and Industry, in its report published in 1931, was conscious of the fact that monetary policy had not accorded fully with the needs of the domestic economy, and noted especially that business found it difficult to interpret the meaning of a change in Bank rate in any particular instance. 'They [the businessmen] may be uncertain whether it merely represents an effort to correct a temporary maladjustment in the international short-term loan position or whether it is the beginning of a contraction which will produce a curtailment of enterprise and business losses until the necessary adjustments have been effected.'[33] It was a well-known fact that the business community regarded Bank rate as an important psychological barometer but then the Bank had never disguised the fact that external considerations took priority in determining changes in its discount rate. In fact before 1914 the rate weapon had been used much more vigorously to protect the Bank's reserves; between 1900 and 1914 Bank rate was altered nearly 70 times. By contrast, in the postwar period it was kept much more stable; the average frequency of change was about half that of prewar and the rate often remained stable for lengthy periods of time. Moreover, though the Governor of the Bank of England (Montagu Norman) did not consider monetary policy as an instrument designed specifically for the purpose of achieving domestic stability, the Bank did take more account of internal considerations than it had done before 1914, while the Treasury had its own reasons, namely the cost of debt services, for trying to restrain the rise in interest rates. Thus in practice the discount rate was raised to a level no higher than absolutely essential to safeguard the reserves, and to offset the deflationary impact open-market operations were often employed. In fact policy was aimed at producing a minimum disturbance to the cash basis of the banking system and as a result the overall liquidity situation of financial institutions remained relatively satisfactory in the later 1920s.

The frequently voiced criticism of contemporary industrialists was about the cost and availability of credit. Historically high interest rates at a time of falling prices was not the most conducive climate for investment. High interest rates also made non-industrial investment such as gilt-edged securities more attractive as an outlet for savings and in some cases firms with spare cash took opportunity of the good yields in preference to more risky investment projects. During the 1920s gilt yields averaged 4.5 per cent and competed strongly for investors' funds thereby making it more difficult for industrialists to secure accommodation at the right price in the open market. Another deterrent was the high cost of servicing debenture loans contracted in the postwar boom by many firms in the staple trades, with little prospect of funding these at lower cost given the prevailing structure of interest rates. Much therefore depended on a firm's or industry's individual circumstances. In cases where expectations were good, as for example in the newer industries, there was probably little difficulty in acquiring finance irrespective of the yield on alternative investments. The upward trend in new issues on the stock market, culminating in the minor boom of 1928-29, would seem to suggest that new developments were not hindered by lack of finance. On the other hand, in the case of the staple industries with their burden of debt and poor growth prospects, the opportunities for raising finance were much more limited, though in this instance it is likely that the burden of excess capacity was a more powerful deterrent to investment than the cost and availability of financial accommodation.

As far as bank advances and overdrafts are concerned it is

possible to be rather more specific on the question of cost. It was generally considered that changes in interest rates within a certain range had only a limited effect on industrialists' ability and willingness to borrow from the banks, provided of course no form of credit rationing was imposed. Norman held the view that Bank rate had little effect on the supply of credit to industry and trade unless it was raised above 4.5 per cent and held there for any length of time. This view was supported by the fact that overdrafts and loans were mde by the commercial banks at a charge of between 0.5 and 1 per cent above Bank rate but with a minimum floor to charges of around 4.5 to 5 per cent, though larger customers and those with a high credit rating might secure more favoured treatment.[34] In practice therefore, a 4.5 per cent Bank rate was the effective cut-off point up to which the charges of the banks would not vary very much except perhaps at times of extremely low Bank rate of around 2 per cent. Now after 1921 average Bank rate remained around or slightly below the critical point for much of the decade, the major exceptions being 1926 and 1929 when it hovered around the 5 per cent mark. Moreover, if we exclude these years then average Bank rate was only slightly higher than it had been in the period 1904-13 (4.15 as against 3.75 per cent). In other words it could hardly be maintained that businessmen were hindered by steep charges for bank accommodation as a consequence of Bank rate policy in this period, though there may have been a real cost effect given the downward slope of the price level.[35]

This conclusion is also supported by the trend in bank advances. Following the sharp contraction after the postwar boom, bank advances rose steadily from 1922 onwards and by the later 1920s the volume of lending was substantially higher than in 1919-20. The banks were able to achieve this, despite the relative stability of the monetary base, by rearranging their asset portfolios. As advances rose investments were reduced so that advances as a proportion of total deposits increased from 45.1 per cent in 1921 to a peak of 55.5 per cent in 1929.

As far as the availability of credit is concerned, one could even argue that it was granted too readily in the case of the staple industries. Porter has shown how the commercial banks placed themselves in an impossible situation because of their generosity in supporting cotton-spinning companies after the collapse of the postwar boom. In an effort to protect their advances they continued to extend lines of credit thereby prolonging the agonies of the industry when 'the logic of the industry's position dictated a reduction in capacity and employment'.[36] Consequently, capacity in the cotton industry failed to respond to market conditions in the 1920s.

While monetary aggregates and policy were clearly deflationary in the early 1920s, it would be difficult to argue the same for the latter half of the decade. At all times the Bank had to give priority to the external side but it was ever mindful of the needs of the domestic economy. Residential construction was probably the sector most affected by high interest rates but this was partly offset by the frequent subsidy programmes.[37] There is no strong evidence however that industry in general was seriously handicapped by the high cost or lack of finance and we doubt whether lower interest rates would have made much difference to industrial investment. Excess capacity, especially in the staple industries, was certainly a much greater deterrent to investment. Moreover the evidence of the 1930s, when industrial investment responded very slowly and weakly to low interest rates, would seem further to confirm this point. Several surveys of this period suggest that business firms were affected more by the availability of finance and the psychological effects of frequent changes in interest rates than by the actual cost of finance.[38] If so, it could be argued that the greater stability of interest rates in the 1920s compared with before the war should have compensated for the slightly higher cost of borrowing.

This is not to deny that alternative macro-policies (monetary, fiscal and exchange rate) would have appeared more appropriate under the domestic conditions of the time. But it is very doubtful whether they would have solved the problems of the 1920s. It was not simply a question of reviving confidence after a cyclical downturn. The main problem was a long-term structural one, as reflected in the patchy nature of development, and this required substantial readjustment in terms of resource allocation. Such adaptation was bound to take time and it was not readily amenable to macro-policies. As Morton noted with reference to discount policy, it 'was never designed to bring about readjustment of great industries and their fixed plants, to rectify maladjustments in world trade, or to compel a nation to adapt itself to vast changes in the world's industrial and commercial structure.'[39] Nor were such policies likely to do much to improve the internal efficiency of Britain's industries. The transition would no doubt have been eased by selective fiscal measures together with regional and industrial policies designed to speed the process of creative adjustment, but such were not forthcoming in this period. Moreover, we emphasise the latter point in particular since, as we shall see, the subsequent record with such policies is not one which commands respect since they have tended to preserve the status quo rather than generate creative adjustment.

Though there was no question, for reasons already discussed, of radical policy responses to the domestic problems of the 1920s, what we have tried to do here is to question whether real economic magnitudes can be easily manipulated by policy action in a situation in which large-scale structural transformation is taking place. The experience of the 1930s offers some interesting evidence as to the forces of recovery and the role of policy and we shall return to these themes in the subsequent chapters.

## Notes

1. K.O. Morgan, *Consensus and Disunity: The Lloyd George Coalition Government, 1918-1922* (1979), pp. 106, 108.
2. Fifty-five per cent of all wage reductions that took place in 1921 and 38 per cent of those in 1922 were a direct result of sliding-scale agreements. Thereafter indexation agreements fell out of favour.
3. In 1921, according to Reader's graphic words, 'the nation suffered a severe heart attack and perhaps it has never truly recovered.' W.J. Reader, *Bowater: A History* (1981), p. 22.
4. C. Barnett, *The Collapse of British Power* (1972), pp. 485-7.
5. The real wage level may be a more important factor in delaying recovery than modern economists allow, though in recent years some have been revamping the pre-Keynesian views on the influence of the real wage level. See M. Casson, *The Economics of Unemployment* (1983). We shall have cause to return to this point again in what we refer to as periods of 'high turbulence' when the real wage level is thrown out of equilibrium leaving the exchange rate and productivity as the safety valves.
6. J.A. Dowie, '1919-20 is in Need of Attention', *Economic History Review*, 28 (August 1975), p. 447.
7. If allowance is made for those outside the insured trades, the percentage unemployment for the decade is lower, 7.5 per cent, though the absolute numbers involved are of course higher.
8. S. Tolliday, 'Tariffs and Steel, 1916-1934: The Politics of Industrial Decline', in J. Turner (ed.), *Businessmen and Politics* (1984). pp. 50-1.
9. W. Lazonick, 'Competition, Specialization and Industrial Decline', *Journal of Economic History*, 41 (March 1981) and 'Industrial Organization and Technical Change: The Decline of the British Cotton Industry', *Business History Review* (1982).
10. See A. Slaven, 'British Shipbuilders: Market Trends and Order Book Patterns between the Wars', *Journal of Transport History*, 3 (September 1982); N. K. Buxton, 'Efficiency and Organization in Scotland's Iron and Steel Industry during the Interwar Period', *Economic History Review*, 29 (February 1976) and 'Entrepreneurial Efficiency in the British Coal Industry', *Economic History Review*, 23 (1970), in which the author stresses demand side factors.
11. W.W. Rostow, *Why the Poor Get Richer and the Rich Slow Down: Essays in the Marshallian Long Period* (1980), p. 92.

12. E. D. McCullum, 'The Problem of the Depressed Areas in Great Britain', *International Labour Review*, 30 (August 1934), p. 137.
13. C. M. Law, *British Regional Development since World War I* (1980), pp. 69, 72, 75.
14. *Report of the Royal Commission on the Distribution of Population*, Cmd 6153 (1940), p. 276.
15. N. K. Buxton, 'Economic Growth in Scotland between the wars: the Role of Production Structure and Rationalisation', *Economic History Review*, 33 (November 1980), p. 554.
16. M. P. Fogarty, *Prospects of the Industrial Areas of Great Britain* (1945), p. 423.
17. In any case it requires some modification in the light of recent work; see, for example, G. C. Peden, 'The "Treasury View" on Public Works and Employment in the Interwar Period', *Economic History Review*, 37 (May 1984).
18. A somewhat higher share in fact than in the 1970s and early 1980s. See *Financial Times*, 24 February 1983.
19. It should however be noted that Winston Churchill displayed 'an unequalled ingenuity ... in producing a balanced budget out of what was on any reasonable reckoning a deficit.' U. K. Hicks, *British Public Finances: Their Structure and Development, 1880-1952* (1954), p. 151.
20. Quoted in A. Cairncross and B. Eichengreen, *Sterling in Decline: The Devaluations of 1931, 1949 and 1967* (1963), p. 42.
21. See D. E. Moggridge, 'Policy in the Crises of 1920 and 1929' in C. P. Kindleberger (ed.), *Financial Crises: Theory, History, Policy* (1982), pp. 175-80 and S. Howson, 'A Dear Money Man? Keynes on Monetary Policy, 1920', *Economic Journal*, 83 (June 1973) and 'The Origins of Dear Money, 1919-1920', *Economic History Review*, 27 (February 1974).
22. The use of the 10 per cent yardstick is not entirely satisfactory since, until recently, the extent of the overvaluation has never been verified quantitatively. The degree of overvaluation depends very much on the price indices used and the coverage of currencies. Redmond's calculations for an effective rate for sterling based on a basket of currencies, indicate an overvaluation ranging from 5-8 per cent using wholesale prices, to as much as 20-25 per cent when retail prices are used. He also finds a clear trend towards diminished overvaluation through time whichever price indices are employed, with wholesale prices recording a slight undervaluation by 1929-30 and retail prices indicating a significant overvaluation. J. Redmond, 'The Sterling Overvaluation in 1925: A Multilateral Approach', *Economic History Review*, 37 (1984), pp. 528-30.
23. D. E. Moggridge, *British monetary Policy: The Norman Conquest of $4.86* (1972), pp. 246-50 and *The Return to Gold: the Formulation of Economic Policy and its Critics* (1969), p. 79.
24. N. H. Dimsdale, 'British Monetary Policy and the Exchange Rate, 1920-1938', *Oxford Economic Papers, Supplement* (1981), p. 323; also J. Foreman-Peck, *A History of the World Economy* (1983), pp. 234, 257.
25. B. W. E. Alford, in R. Floud and D. McCloskey (eds.), *The Economic History of Britain since 1700*, Vol. 2, *1860 to the 1970s* (1981), p. 310.
26. I. Drummond in Floud and McCloskey, *op.cit.*, p. 297.
27. *Ibid.*, Dimsdale, *loc. cit.*, p. 328 comes to a similar conclusion though on p. 322 he appears to attach more weight to the adverse affects of the overvaluation.
28. In a recent note Garside and Hatton make extravagant claims as to the responsiveness of the British economy to a lower exchange rate than the one

adopted in 1925. They maintain that the 'problems of Britain's export industries were severely exacerbated by the overvaluation of the pound', and 'that a lower exchange rate could have had powerful effects, both in raising aggregate demand and directing expansion into depressed staple industries.' Quite apart from the questionable issue as to whether it would have been expedient to revive the ex-growth sectors, the authors wrongly assume a *ceteris paribus* condition and favourable trading elasticities to produce the desired result. Unfortunately they then go on to argue the equally false proposition that the exchange rate was the problem in the 1930s. W. R. Garside and T. J. Hatton, 'Keynesian Policy and British Unemployment in the 1930s', *Economic History Review*, 38 (1985), p. 86.

29. A. Maizels, *Industrial Growth and World Trade* (1963), pp. 509-10.
30. Political and Economic Planning, *Report on International Trade*, Chapter 5.
31. S. Glynn and A. Booth, 'Unemployment in Interwar Britain: A Case for Relearning the Lessons of the 1930s', *Economic History Review*, 36 (August 1983), p. 337.
32. Dimsdale, *loc. cit.*, pp. 327-29.
33. *Report of the Committee on Finance and Industry*, Cmd 3897 (1931), para. 218.
34. W. A. Thomas, *The Finance of British Industry, 1918-1976* (1978), p. 58, quotes a minimum of 5 per cent though large borrowers with good collateral could secure loans at 0.5 per cent above Bank rate with a minimum of 4.5 per cent.
35. A point emphasised by R. E. Catterall, 'Attitudes to and the Impact of British Monetary Policy in the 1920s', *International Review of the History of Banking*, 12 (1976).
36. J. H. Porter, 'The Commercial Banks and the Financial Problems of the English Cotton Industry, 1919-1939', *International Review of the History of Banking*, 9 (1974), pp. 15-16.
37. Rent restriction was probably an equally important factor holding back house construction.
38. A series of studies carried out in the 1930s revealed that in general interest rates had comparatively little influence on businessmen's investment decisions. See J. F. Ebersole, 'The Influence of Interest Rates upon Entrepreneurial Decisions in Business: A Case Study', *Harvard Business Review*, 17 (1938-9), and the papers by Henderson, Andrews and Meade in *Oxford Economic Papers*, 1 (1938), 3 (1940).
39. W. A. Morton, *British Finance, 1930-1940* (1943), p. 112.

# 2 Depression and Recovery, 1929-1939: An Overview

The 1930s are often regarded as being the worst years of the interwar period. This impression is understandable for several reasons: the massive worldwide downswing in economic activity and international trade in the early part of the decade; the dramatic fall in prices; the persistently high level of unemployment and the steadily deteriorating international political climate which eventually culminated in war. During the 1920s there had always been a glimmer of hope that things really would get back to normal and, in the United States at least, the strong boom of these years had nourished great optimism about the future. By the early 1930s, on the other hand, the illusion of any return to normality had been shattered and few people held out much hope for the future. Yet as far as the British economy was concerned things turned out rather better, comparatively speaking, than in the 1920s. By international standards Britain experienced a relatively mild recession and escaped the severe institutional financial crises that hit many other countries; the subsequent recovery was also rather better than that abroad; real incomes rose steadily and consumption patterns testify to an improvement in the standard of living. For the majority in work it was a period of material progress and satisfaction. But the old problems persisted and often in a more acute form: unemployment, for example, remained obstinately high even at the peak of the cyclical recovery and poverty was endemic in some parts of the country. Furthermore, while significant structural change had taken place on the industrial front, the problems of the old staple industries and the depressed regions continued to leave a nasty blot on the economic landscape. And despite noticeable shifts in policy direction, the government proved impotent in dealing with the fall-out from market forces.

## Depression and Financial Crisis, 1929-32

The worldwide severity of the depression of the early 1930s is not in doubt. The large falls in income and output, the collapse of trade and the dramatic financial events of the period have little equal either before or since, at least until the 1970s. What is often forgotten however is that in terms of output and income Britain experienced a fairly modest depression — the amplitude was far less than 1921 for example — and had it not been for the very high unemployment and the drama surrounding her sudden departure from the gold standard in September 1931, the events of these years might not have captured the popular imagination to the extent they did both then and since.

On first reflection the mildness of the recession may appear surprising given Britain's obvious export vulnerability and the muted nature of her economic performance in the later 1920s. In one sense of course the weak boom of the later 1920s may have helped to soften the impact of the downturn in as much as the economy had not overshot itself leaving exhausted sectors in its wake as in the United States, so that the chances of a milder recession and an earlier recovery were somewhat greater than in the latter case. On the other hand, given Britain's severe structural problems and her export exposure, there must clearly have been other factors responsible for maintaining a higher floor to real activity than was the case in other countries. Exports took a battering during the depression and the weakness in this sector was apparent before the American economy pulled the world into recession. In Britain there were already signs of impending downturn in the later 1920s irrespective of what was happening across the Atlantic. The crucial factor was Britain's export vulnerability, reflected in particular by her dependence on staple products and the markets of primary producing countries where incomes were anything but buoyant. When the incomes of these countries began to sag in the later 1920s, as a result of the decline in primary product prices, so the demand for British exports began to weaken. Some years ago Corner drew attention to the fact that the demand for British exports from South America, British India, South Africa, the British Colonies and the Far East, which together absorbed some 40 per cent of the country's exports, was beginning to deteriorate in the latter half of 1928, that is some six to nine months prior to the collapse of the American economy.[1]

Britain therefore may well have experienced a recession independent of American influence. But the subsequent downturn in the US economy greatly aggravated and accelerated the deflationary tendencies in this country and elsewhere. The detailed causes behind the American contraction need not concern us here expect to note that it was very much an internally generated collapse, in part a reaction to the violent boom of the later 1920s which resulted in a saturation or partial saturation of certain market sectors, construction and consumer durables in particular, and hence a drying up of investment opportunities.[2] Demand prospects of these sectors were weakened by the skewed nature of real income gains in the later 1920s which limited the scope for further market penetration. The situation was not helped by excessive speculation in financial circles and the weight of buying on credit, which eventually led to policy tightening and the famous stock market crash of October 1929. Since America was such an important element in the world economy other countries were soon affected adversely by the decline in activity there. US import demand fell sharply and this served to depress prices and incomes in exporting countries, especially primary producing countries which accounted for around one half of the American import bill. This in turn created balance of payments problems in these countries and thereby reduced their ability to import from industrial countries including Britain. At the same time the flow of American credit was being curtailed (the initial break in foreign lending in fact preceded the downturn in activity), which created severe problems in many debtor countries, notably in Central and Eastern Europe and in Latin America. It compounded the already precarious balance of payments position in these countries and eventually precipitated the international financial crisis of the summer of 1931. In short, given the sheer size of the US economy and the unstable and weak financial position of many debtor countries dependent upon it, it is scarcely surprising that the course of events in the United States had a profound effect on the health of the international economy.

For Britain probably the worst features of the depression were the sharp deterioration in the external account and the high level of unemployment. Given the export-orientated nature of British industry and its inherent weaknesses, it was perhaps only to be expected that this would be the case. The volume of exports fell by no less than 37.5 per cent between 1929-32, though even this should be put in perspective by recalling the 30 per cent fall in a single year, that of 1921. Income from invisibles also declined sharply whereas import volumes were maintained. In fact changes in the net trade position more than accounted for the absolute decline in GDP volume between 1929-31, with consumption making a positive contribution.[3] The upshot was that the balance of payments on current account moved from a good surplus of £104 million in 1928 to a deficit of £114 million in 1931, an important factor in the sterling crisis of that year (see below).

The second alarming feature was the very sharp rise in unemployment, much of it occasioned by the collapse in exports of the old staple industries. It rose from a low of nearly 1.2 million, or 9.5 per cent of the insured workforce, in the second quarter of 1929 to around 3 million (22.7 per cent) in the third quarter of 1932, a figure which underestimates the absolute numbers involved because of the exclusion of the non-insured and the effects of short-time working. What made the unemployment problem seem even more glaring was the fact that it was the regions dependent on the export-orientated staple trades, such as coal, textiles, shipbuilding and iron and steel, that bore the brunt of the recession, though numbers out of work rose everywhere. Shipbuilding, for example, literally collapsed, with insignificant launchings at the trough of the depression, so that unemployment averaged nearly 60 per cent in 1932, a rate two and one half times greater than in 1929. The other major staple trades also recorded massive unemployment rates, iron and steel 48.5 per cent, coal 41.2 per cent and cotton 31.1 per cent, as against an average for all industries of nearly 23 per cent (1932). Coal output dropped by a fifth, while cotton exports were halved in two years, largely due to the collapse of the Indian market. Since these sectors of activity were geographically concentrated, in the North of England, in Wales and in Scotland, it was these regions that were badly hit. They had unemployment rates of 30 per cent or more in 1932, that is, over twice those of London and the South-east. Moreover, some areas or localities with excessive dependence on the staple industries had unemployment rates of two thirds or more, for example, Jarrow, Merthyr, Dowlais and Brynmaer, a striking contrast to the low rates recorded in some places in the Midlands and South-east as in the case of Oxford 5.1 per cent, St. Albans 3.9 per cent and Coventry 5.1 per cent (1934). Even by 1936, after several years of recovery, rates of unemployment still remained very high in the North of the country, and over two thirds of the registered unemployed were to be found in Scotland, Wales, Northern Ireland and Northern England.

The data given so far serve to emphasise the worst aspects of the depression, but many of the aggregate indicators of activity tell a less depressing story. While unemployment was exceptionally high, in part aggravated by a strong rise in the labour force during this period, it should be set against the quite modest fall of 4.7 per cent in total employment. Apart from investment, which declined by 14 per cent and mostly in the last year of the depression, most other main indicators recorded moderate falls. The decline in real income and domestic output was of the order of five per cent or less, while consumers' expenditure fell only in the last year of depression and actually rose slightly through the years 1929-32. Even industrial production fell by much less than it had done in 1921, 11.4 as against 18.6 per cent. Moreover, when set against the trends in other countries, the British record appears quite mild. The United States, for example, suffered declines in real income, industrial production and employment of one third or more. If few countries could match the real losses of the American economy, equally few could compare with the mildness of the British recession as the figures in Table 1 demonstrate:

*Table 1: Changes in GDP and Industrial Production in Selected Countries, 1929-32*

| | GDP | Industrial Production |
|---|---|---|
| United States | -28.0 | -44.7 |
| Austria | -19.8 | -34.3 |
| France | -11.0 | -25.6 |
| Germany | -15.8 | -40.8 |
| Sweden | -4.0 | -11.8 |
| UK | -5.1 | -11.4 |

*Sources:* A. Maddison, *Phases of Capitalist Development* (1982), pp. 174-5; OEEC, *Industrial Statistics, 1900-1959* (1960), p. 9.

As already intimated, there were of course considerable variations in the amplitude and timing of the swings in activity among different sectors of the economy. While industrial production and transport and communications declined by 11.4 and 8.7 per cent respectively, the output of services and distribution increased slightly. Export-sensitive industries, as one might expect, tended to be first in and last out of depression and experienced swings of large amplitude, whereas consumer goods industries (especially those producing non-durables) lagged at the upper turning point and responded early in the recovery phase and had lower amplitude. Producer goods industries, on the other hand, lagged at both turning points and experienced large swings. Thus by far the largest contractions were recorded in shipbuilding, mechanical engineering, ferrous and non-ferrous metals, whereas movements in chemicals, tobacco, drink, clothing and textiles were more modest. In some industries the depression was very limited, being confined to one year only. Electrical engineering, for example, declined in 1931 only, but over the course of the recession its output actually increased by five per cent. Food manufacture, paper and printing and electricity generation emerged more or less unscathed, with output increasing in each case. In clothing, the contraction was very modest, while the leather industry recorded a significant increase in output largely because there had been a check to activity a year before the slump. These examples serve to show that even in a significant recession there is a wide range of variation between different sectors of the economy, and that growth industries are able to maintain progress even in the most difficult times.

From the authorities' point of view probably the most traumatic aspects of the period were the events leading up to the abandonment of the gold standard in September 1931. The Labour Government of 1929, no less than the previous Conservative one, was committed to retaining the existing parity of sterling. In the course of time, however, it became an increasingly hollow commitment given the pressure of international events, the weakness of Britain's external position and the inevitable conflict in policy terms between the domestic economy and the external position. The detailed events leading up to the international financial crisis of the summer of 1931 have been the subject of several studies and only the features of special relevance to the British position need be recorded here.[4] On the international side the beginning of the pressure on international liquidity really began in the late 1920s, when primary producing countries were faced with falling prices and incomes due to overproduction and stockpiling, and at the same time creditor nations began to curtail their lending abroad. Agricultural prices fell by about one half between late 1929 and the middle of 1931 and many primary producing countries, mostly debtors, were faced with large payments deficits and an outflow of funds. Demand for accommodation from the main creditor countries was not forthcoming however. In fact the flow of funds, both on long- and short-term accounts, from the two main creditor nations, the US and France, slackened significantly in 1928-29, and by 1930 both countries had become capital importers. The other major creditor nation, Britain, was not in a strong enough position financially at this time to render assistance on a sufficient scale. In the absence of accommodation from the creditor countries, several primary producing countries were forced to leave the gold standard and devalue their currencies. By 1930, six Latin American countries, Canada, Australia and New Zealand had abandoned gold.

It is possible that had the creditor nations been more willing to assist the primary producers at the onset of their difficulties the magnitude of the subsequent financial crisis might have been lessened. But by the spring of 1930 the situation was virtually beyond repair. By then the deflationary process had taken hold firmly in Europe and the United States and with falling incomes, unbalanced budgets, balance of payments deficits and internal monetary difficulties, the reaction of governments was one of retrenchment and deflationary policies which only tended to aggravate the situation. In these circumstances there was little prospect of debtor countries securing the accommodation required. Moreover, during 1930-31 the international financial situation was exacerbated by institutional weaknesses as a wave of bank failures spread across both the United States and Europe. In America alone 1,345 banks collapsed in 1930 and a further 687 folded in the first half of 1931, many of them heavily involved in agricultural operations. During the latter part of 1930 and in 1931 the banking crisis spread to Europe, culminating in the failure of the Austrian Kredit Anstalt and the collapse of the German banking system in the spring and summer of 1931. The main effect was to generate a scramble for liquidity and much of the strain fell on London as one of the few centres still willing to grant accommodation. Between June 1930 and December 1931 London lost £350 million of foreign funds and the outflow reached panic proportions in the summer of 1931 when something like £200 million left the country. This rapid rate of withdrawal was largely occasioned by the continental financial panic, but the position was undoubtedly aggravated by the loss of confidence in Britain's ability to maintain solvency as her external payments position deteriorated and in the light of her unfavourable short-term liquidity position.[5] In the circumstances there seemed to be little else to do but to release the sterling parity, and so on 21 September 1931 Britain finally abandoned the gold standard and devalued her currency.

On reflection there was probably very little prospect of Britain retaining the sterling parity let alone containing the international financial crisis given her weak external position and the precarious domestic budgetary situation. Within three years the current account surplus of over £100 million in 1928 had been converted into a deficit of roughly equal magnitude. The deterioration of the trade balance due to the sharp decline in exports was only partly responsible for the turnround since it was offset somewhat by an improvement in the net barter terms of trade. The main problem was the severe contraction in invisible earnings arising from the fall in returns on overseas investments as interest rates fell, and the decline in income from shipping and commercial and financial services as world trade contraction reduced the services required by foreigners. Between 1929 and 1931 there was an adverse swing in Britain's invisible balance of over £130 million, that is, approximately twice the magnitude of the deterioration in the trade balance.[6]

The second main problem was Britain's short-term liquidity position. It is now apparent that Britain's short-term creditor position was much weaker than even contemporaries realised. Before the war her liquid assets (mainly gold and sterling acceptances on foreign account) had been well in excess of her short-term liabilities. A contraction of between £250 and £300 million in her stock of short-term assets left Britain a net debtor on short-term account at the end of the war. During the 1920s short-term liabilities tended to increase, but short-term liquid assets did not rise correspondingly, while some of the short-term funds were used for long-term investment and were therefore difficult to raise at short notice. Thus by the middle of 1930 short-term liabilities were more than twice liquid assets. The precise position may never be known, but it appears that gross liabilities totalled about £760 million, which could be offset by short-term foreign assets of £176 million and gold reserves of £156 million, giving a net short-term deficit of around £430 million. As long as confidence in London's financial position remained secure this debtor status on short-term account posed no special problem. But once the demand for liquidity became acute and foreign balances were withdrawn as in 1930-31, it soon became apparent that London could not withstand the strain for long without soon depleting her reserves, realising her short-term assets or further borrowing abroad. All these alternatives presented certain problems. The rapid depletion of the gold reserves tended to reduce confidence further and would probably have led to an earlier collapse in sterling. In any case, the Bank of

England's reserves (mainly gold and foreign exchange) were only sufficient to cover a fraction of the liabilities since part of the reserves was earmarked as a backing for the domestic note issue, while recourse to special measures to release additional reserves to support the currency were more likely to have aggravated than alleviated the market's fear and uncertainties about Britain's precarious financial state. The second line of defence, that of realising Britain's short-term assets abroad, was rendered difficult owing to the fact that many of these were locked up in European financial centres which were themselves in an illiquid state. Effectively therefore many of them were frozen during the course of the crisis. The only other immediate solution was to borrow abroad to cover the outflow of funds. But since many countries were experiencing acute financial difficulties at the same time this was easier said than done, though in the final stages of the crisis (August-September) some £130 million worth of credits were scraped up in New York and Paris. 'It was this inability to either retain previously invested funds in London or to attract further investment by foreigners that was the root cause of London's short-term difficulties.'[7]

It has been suggested that throughout the crisis the Bank of England's policy was weak and ineffective and that this only served to undermine confidence in London's position. No attempt, for instance, was made until late in the day to raise intereest rates in order to staunch the flow of funds from London and attract an inflow from abroad. Williams suggests that had the Bank raised its discount rate early in 1931 some of the pressure might have been taken off London. It is doubtful however whether this alone would have solved the underlying problem. The loss of confidence and severity of the liquidity panic were so great that it is unlikely that much could have been done to retrieve the situation. According to Clarke, sterling was beyond salvation whatever measures had been adopted.[8] In any case, raising the Bank rate might well have been counter-productive in the sense that it could have further increased the distrust in sterling. Moreover, at the critical period the loss of confidence was heightened by the alarming budgetary situation. Britain's growing budgetary deficit had already begun to give grounds for concern in foreign circles by the May of 1931, and by August of the same year it had reached a state of alarm following the gloomy prognostications of the future outturn of the budget made by the May committee in their report on National Expenditure published at the end of July. The May committee forecast a budget deficit of no less than £120 million for 1932 and called for severe economy measures to rectify the imbalance. Thus the budgetary situation became a major factor in affecting sentiment in Britain's financial soundness. Not surprisingly therefore, when the Bank did finally raise its discount rate to 4½ per cent late in July it had little impact in terms of restoring confidence.

It still remains something of a mystery however why the Bank did not resort more readily to the traditional remedy of dealing with a drain on its reserves. It is possible, as Cairncross and Eichengreen suggest, that the Bank withheld further increases in its rate in an effort to force the government to balance its budget.[9] A second possibility is that in the turbulent conditions of the time the Bank had doubts about the efficacy of the rate weapon. On the other hand, the most generally accepted explanation is that it was the awareness that higher interest rates would make the domestic economic situation even worse which prevented the Bank from taking a stronger line. During the course of the 1920s the use of the interest-rate weapon as a flexible instrument to control the exchanges had been circumscribed in part by domestic considerations, and with the worsening economic conditions after 1929 it was becoming clear that the twin goals of domestic and external stability could not be pursued simultaneously. One had to be abandoned and in the end it was the gold standard. Devaluation, however distasteful to the authorities, was preferable to further doses of domestic deflation to secure adjustment under the fixed exchange rate regime of the gold standard. Perhaps the one slight snag about this argument is that when the gold standard was finally abandoned Bank rate was hoisted to 6 per cent as a gesture of reassurance and to prevent inflation!

## Aftermath of the Crisis and Shift in Policy Direction

The financial crisis of 1931 and Britain's departure from gold marked a turning point for the international as well as the British economy. Most countries, though with a few important exceptions such as France, Belgium, Germany and Poland, followed Britain by relinquishing the gold standard and devaluing their currencies, culminating with the dollar in 1933. This action paved the way for a shift in policy emphasis towards domestic recovery measures, since countries were no longer constrained by the financial dictates of the gold-standard regime. On the other hand, breaking the link with gold did not automatically guarantee that external influences would not jeopardise the process of recovery. While it left greater room for manoeuvre on the domestic policy front, particularly in the case of countries securing an initial advantage by devaluing early, it could not in the long run remove the possibility of external disequilibrium arising and upsetting any attempt to stimulate internal recovery. To counter this prospect there was a widespread move to shield domestic economies from possible adverse influences on the external side by a battery of restrictions including higher tariffs, import quotas, exchange controls, capital embargoes and special devices to iron out fluctuations in exchange rates. Barter trading deals and agreements also became common practice, partly as a response to the much more restrictive trade policy. Thus the pre-1914 system of multilateral trade and payments and the free flow of commodities, capital and labour across national boundaries, on which the gold and sterling monetary system had rested, finally came to an end. Instead, nationalistic economic policies, managed currencies, restrictive trading policies and bilateral trade deals became the order of the day, all of which severely limited the scope for any significant recovery in world trade following the sharp collapse in the depression. The finishing touch to the old regime came with the breakdown of the World Economic Conference in 1933 which effectively 'signalised the end of any general attempts at international action in the economic field during the inter-war period.'[10]

British policy in the 1930s, though marking a significant departure with past practice, was not nearly so autarchic as in some countries, for example, Germany and Eastern European countries, but the pattern of development, especially on the external side, was quite similar. Sterling was devalued in 1931, an embargo was placed on foreign lending, tariffs and, to a lesser extent, quotas were employed to shut out imports. Bilateral trade agreements and imperial preference trading arrangements became widespread, while fluctuations in the exchange rate were taken care of by the establishment of the Exchange Equalisation Account in April 1932. Most of the changes in external policy had been implemented by the end of 1932, though the effects on the beginnings of recovery were mixed. The shift in domestic policy was far less radical. It was confined largely to the provision of cheap money and *ad hoc* measures designed to assist specific sectors of the economy, for example, industrial reconstruction schemes, agricultural marketing devices and special areas legislation, the impact of which was fairly limited. By contrast, fiscal policy remained remarkably orthodox following the traumas of the public expenditure crisis in 1931 and therefore provided no leverage to the economy.

Despite the flurry of policy activity in the early 1930s, by and large it was natural forces rather than policy factors which were responsible for Britain's recovery. More specifically, it was a domestically based recovery which derived little strength from the external side. Some of the policy measures, as we shall see, did provide a measure of assistance though not all of them were aimed specifically at recovery or employment, while some were mutually contradictory. Before looking at the role of policy factors and the forces of recovery in more detail we turn first to a brief outline of the recovery process.

## A Review of the Recovery Process

Most writers would agree that Britain experienced a fairly strong and sustained recovery in the 1930s and that, compared with her relatively poor showing in the previous decade, she did rather better than many of her leading competitors. In fact for the only time in the twentieth century Britain secured a position in the top half of the growth league table. There is rather less agreement, however, as to the factors responsible for recovery, particularly as between policy induced and natural forces of recovery, and the contribution of different sectors of the economy.

The main force of the depression had been spent by the end of 1931 though it was not until the latter half of 1932 that signs of recovery became apparent. Unemployment peaked in the third quarter of that year and most economic indicators continued to decline or stagnate for much of the year. The lower turning point can probably be located somewhere in the third quarter of 1932 since by the final quarter the economy generally was showing signs of picking up. The beginnings of recovery were somewhat earlier than in most other countries, though the quarterly and monthly data for several indices suggest that it was still weakly based.[11] It was not until the following year, however, that most indicators of activity registered a marked improvement so that 1933 can be taken as the first full year of recovery. It is true that there had been signs of revival in some of the export-sensitive industries such as textiles late in 1931, following the devaluation of the currency. But this was a very limited and temporary affair and in no sense could it be argued that export-led growth was the key to the initial upturn from the slump. The total volume of exports, which had fallen by close on 40 per cent during the depression, remained stable in 1932 and rose by only just over one per cent in the following year, while many of the heavy export-orientated trades such as mining, shipbuilding, and mechanical engineering remained very depressed. Exports of coal, for example, were flat between 1932 and 1935. Instead the first real thrust to recovery came from non-export sectors such as building and related trades, transport (mainly road transport), electricity and the newer sectors of manufacturing such as vehicles, electrical engineering and the consumer durable trades. By 1934 sustained growth had extended to many sectors of the economy though it tended to be most vigorous in the new and domestically based industries and the industries associated with these sectors, and least prominent in the old exporting staple trades. Nevertheless, exports, apart from coal, did pick up quite smartly in 1934 and 1935 and this, together with derived demand from the newer industries and building for metal inputs, helped to bring a measure of recovery to the older industries. Investment also rose strongly in 1934 when the largest increase of the interwar years was recorded. The recovery in employment was more modest however and by the end of 1934 there were still over two million workers unemployed or nearly 16 per cent of the insured population.

During the course of the next three years the pace of recovery continued practically unchecked. Virtually all industries expanded their output during this period, though the boom was primarily a domestic one still, the leading sectors being construction, transport, the newer industries including consumer durables and certain capital goods industries. Most non-durable consumer goods industries and services also grew but their rate of expansion was below the general average, while the rate of growth in agriculture was very low indeed. Exports continued to expand modestly but they remained well below pre-depression levels even by 1937. The process of recovery during these middle years was assisted by the revival of the world economy and international trade and also by the beginnings of rearmament.

By the middle of 1937 there were signs that the boom was coming to an end. The rate of growth of incomes and output slowed down during the course of the year and consumption was restricted by the unfavourable trend in real wages. The latter were more or less stable between 1934-35 and then declined in the following year. The boom was reflected in rising prices and costs, a trend accentuated by the increasing pressure on international commodity prices as recovery gathered pace abroad. Wholesale prices, for example, rose by no less than 15 per cent in 1937. Moreover, in some areas and trades — engineering, construction and the new sectors in the Midlands and South — shortages of labour, particularly skilled labour, were beginning to make themselves felt despite the still high level of unemployment. There were indications too of some drying up of investment opportunities, especially in housing and consumer durables. These unfavourable trends, coming as they did after nearly five years of uninterrupted growth, were bound to dampen optimism. Thus some slowing down in the recovery was probably inevitable, though the position was undoubtedly aggravated at the time by the rather sharp drop in exports in 1937-38 following the downswing in activity in the United States.[12] Even so, the recession in Britain was relatively mild and short-lived. Industrial production and domestic output declined by one or two per cent, while real incomes and consumption continued to expand modestly. The favourable shift in the terms of trade in 1938 was mainly responsible for helping to maintain the latter. Few industries suffered severe declines in output except for those dependent on exports, such as shipbuilding, textiles, ferrous metals and coal.[13] By the beginning of 1939 the economy had resumed its upward trend being backed now by rearmament and later by war.

Overall Britain's record of achievement throughout the recovery of the 1930s was quite impressive. Between 1932 and the peak in activity in 1937 real income increased by 19 per cent, gross domestic production by 23 per cent, industrial production by nearly 46 per cent, gross fixed investment by 47 per cent, and even exports by 28 per cent. Moreover, in absolute terms the level of economic activity was far higher than it had been in 1929 or 1913, the major exception being the volume of exports which by 1937 was still 35 per cent below the pre war level and 20 per cent less than the already depressed volume of 1929. But unemployment and the associated regional problems remained black spots. Despite the considerable upsurge in activity total employment rose by less than 17 per cent, a reflection in part of the steady increase in productivity. Consequently unemployment did not fall as far as might have been expected. Even at the peak of recovery, in the third quarter of 1937, the total unemployed still numbered 1.4 million or just over 9 per cent of the insured population. However, it should be noted that the greater part of this consisted of structural unemployment and that most of the cyclical unemployment had been eliminated.[14] Thus by the later 1930s a more or less full cyclical recovery had been achieved.

In fact the years in question witnessed the largest and most sustained period of growth in the whole of the interwar period, and of the first half of the twentieth century. Internationally Britain also fared much better than she had done previously or was to do subsequently. Although a few countries, such as Germany and the United States, registered rather higher rates of growth between 1932-37 due to sharper contractions in the depression, if we take as a basis of comparison the more meaningful period between cyclical peaks, 1929-37, then Britain's performance shows up very well indeed. Whereas in the 1920s Britain lagged behind most other major countries and hence lost heavily in terms of trade and output shares, this trend was partly reversed in the 1930s as a result of the relative mildness of Britain's recession and the strength of the subsequent recovery.

Thus, several countries, notably France, Belgium, Canada and the United States, failed to regain or surpass their 1929 levels of activity, while only four countries, Germany, Norway, Denmark and Sweden, recorded a better performance than Britain in total output and industrial production over the period 1929-37. In fact this is probably the only period in the twentieth century when Britain outpaced her main competitors. Even in exports, where volume fell short of the 1929 level by a large margin, there were several countries with an inferior outturn, whereas in the 1920s (or 1913-29) Britain's trading performance was the worst on record. Consequently, Britain's share of world trade and output stabilised in the 1930s. Her share of world manufacturing production remained at about the same level as in the late 1920s, at just over 9 per cent, while her share of world manufacturing exports declined slightly from 23.6 per cent in 1929 to 22.4 per cent in 1937 as against nearly 30 per cent in 1913. In terms of total world trade, Britain's share also remained fairly stable in the 1930s at around 10 per cent.

The recovery was accompanied by several important structural changes in the economy, though most of these merely represented the continuation of trends apparent after the first world war. Since the recovery was primarily a home market one, production for export declined in importance; in 1913 it had accounted for nearly one third of total industrial production, declining to 22 per cent in 1930 and to 15 per cent by 1938. The large export-orientated staple trades bore the brunt of this contraction (see Chapter 1), hence the big shift of labour out of these sectors and the emergence of the regional problem. The importance of these industries, both in terms of output and employment, continued to dwindle during the 1930s, though perhaps not so rapidly as previously, due to protective policies and later on the effects of rearmament. By contrast, it was the newer industries, motor manufacturing, electrical engineering, rayon, chemicals, road transport etc., together with the building trades and many service sectors, much of whose activity was concentrated in the southern half of the country, which accounted for an increasing share of output, employment and investment. The shifting pattern of activity in the industrial sector can be seen from the following figures on new and staple industry sector shares:

*Table 2: New and Staple Industries as a percentage of Net Industrial Output of all Census of Production Trades*

| | 1924 | 1930 | 1935 |
|---|---|---|---|
| New industries | 14.1 | 15.9 | 21.0 |
| Staple industries | 37.0 | 29.6 | 27.8 |

*Note:* Staple industries include mining and quarrying, iron and steel, shipbuilding, mechanical engineering and cotton and woollen textiles. New industries comprise vehicle manufacture, electrical engineering, rayon, non-ferrous metals, paper, printing and publishing.

*Source:* N. K. Buxton, 'Economic Growth in Scotland between the Wars: The Role of Production Structure and Rationalisation', *Economic History Review*, 33 (1980), p. 549.

The role of the new industries and that of building in the recovery process has been the subject of renewed debate in recent years and we shall return to this issue in Chapter 6, which examines the structural elements of recovery in more detail.

## Notes

1. D.C. Corner, 'Exports and the British Trade Cycle, 1929', *The Manchester School*, 24 (1956), pp. 132-6.
2. The causes of the US depression have been the subject of lively debate in recent years. There is a good account in R. A. Gordon, *Business Fluctuations* (1952), Chapter 13.
3. It is interesting to compare the position with that of 1979-81 when the net trade position was slightly positive and most of the GDP contraction was due to the sharp declines in stockbuilding and fixed investment. For instructive

comparisons see G.D.N. Worswick, 'Two Great Recessions: the 1980s and the 1930s in Britain', *Scottish Journal of Political Economy*, 31 (1984), pp.211, 223.

4. See for example D. Williams, 'The 1931 Financial Crisis', *Yorkshire Bulletin of Economic and Social Research*, 15 (1963) and 'London and the 1931 Financial Crisis', *Economic History Review*, 15 (1962-3); also A. Cairncross and B. Eichengreen, *Stirling in Decline* (1983), pp.44-82.
5. See B. Eighengreen, 'Bank Failures, Balance of Payments and the 1931 Sterling Crisis', *Harvard Institute of Economic Research Discussion Paper*, No 869 (Dec. 1981).
6. Cairncross and Eichengreen, *op. cit.*, pp.35, 37; Eichengreen (1981), *loc. cit.*, p.5; D. Moggridge, 'The 1931 Financial Crisis — A New View', *The Banker*, 120 (August 1970), pp.332-9, who stresses the role of Britain's deteriorating invisible balance in the crisis of the gold standard.
7. Williams, *Economic History Review*, *loc. cit.*, p.522.
8. S.V.O. Clarke, *Central Bank Cooperation, 1924-1931* (1967), p.204.
9. Cairncross and Eichengreen, *op. cit.*, p.64.
10. H.W. Arndt, *The Economic Lessons of the Nineteen-Thirties* (1944), p.242.
11. Monthly and quarterly data are available in F. Capie and M. Collins, *The Inter-war British Economy: A Statistical Abstract* (1983). For progress in other countries see League of Nations, *World Economic Survey 1932-3* (1933).
12. Richardson maintains that the recession, which started in the autumn of 1937, was due as much to internal causes as to the fall in activity in the United States. See H.W. Richardson, 'The Economic Significance of the Depression in Britain', *Journal of Contemporary History* 4 (1969), pp.7-8. A more detailed analysis can be found in the same author's book, *Economic Recovery in Britain 1932-39* (1967).
13. The major exception to this category was the leather industry with a decline in output of 12.6 per cent between 1937-8.
14. Despite the strength of the recovery doubts have been expressed about its resilience following the downturn of 1937-38 and the continued high level of unemployment. Capie and Collins maintain that the economy was only saved by the coming of war. Unfortunately they somewhat exaggerate the extent of the recession of the late 1930s and fail to appreciate that one can have full cyclical recovery below the full employment level. F. Capie and M. Collins, 'The Extent of British Economic Recovery in the 1930s', *Economy and Society*, 23 (1980).

# 3 Economic Policy and Recovery: External Policy

The principal task now is to determine the main factors influencing Britain's recovery in the 1930s. This promises to be a lengthy undertaking because since the present writer first began work on the interwar years there has been an enormous amount of new work, with revisions and counter-revisions to conventional wisdom. However, it also promises to be a worthwhile effort since many of the issues are not without relevance to the current situation in the 1980s.

The main area of debate centres on the question as to whether the recovery was spontaneous or natural as opposed to being policy-induced. It seems appropriate therefore to look first at the role of policy in order to determine the answer to the second part of the question. As already indicated, there was a fairly large number of policy changes in the 1930s all of which could have some bearing on recovery. However, it should be stressed that the policy changes of the period did not form part of a coherent package designed specifically with the intention of promoting recovery and employment. Many of them were adopted in an *ad hoc* fashion in response to the pressure of events, though by the spring of 1932, when the spectre of inflation arising from currency depreciation had not materialised, the Treasury opted in favour of cheap money and a low pound as desirable on recovery and employment grounds. Nor was there anything akin to a major fiscal boost of the sort we have become familiar with in the post-war era, at least until recently. As it happened, however, most of the policy changes of the period did have recovery potential and once the immediate crisis had passed the government became anxious to provide the 'right' climate for the revivial of private enterprise. Thus, if to latter-day Keynesians the direction of policy fell short of the ideal in the absence of a significant fiscal thrust, the government's fiscal policy was not deemed to be entirely negative to contemporaries since the effort made to balance the budget had an unspecified confidence effect.

The range of policy instruments is quite large but for convenience of analysis they may be grouped into three main categories: (1) External policy comprising primarily devaluation and tariff protection including new trading arrangements; (2) Macroeconomic policy which covers cheap money and fiscal policy; and (3) Industrial and regional measures of assistance. Though fiscal policy was a non-starter for much of the period as far as employment creation was concerned, it has greater relevance later in the period via the rearmament programme. In addition, there are several interesting questions which might be posed with regard to fiscal policy, namely the confidence factor of a balanced budget, the reasons why the National Government did not resort to fiscal reflation, and the viability of fiscal reflation in terms of the economic background of the time.

First however we turn to a discussion of external policy. The two major policy changes on the external side were the devaluation of sterling and the imposition of general tariff protection together with associated trading arrangements. There was also a temporary embargo on overseas capital issues following the abandonment of the gold standard, and non-sterling issues remained subject to control in subsequent years. The potential ramifications of this shift in policy were substantial, though in practice the impact was somewhat disappointing from the point of view of the contribution to recovery. The precise effects are also difficult to unravel, especially on the import side where devaluation and the tariff worked in the same direction.

## Devaluation of Sterling

We begin first with the potential effects of a devaluation. Normally one would expect a devaluation of a currency to improve a country's economic health providing certain conditions hold. Devaluation should improve competitiveness in international trade as export prices fall in terms of foreign currency, while import prices rise in terms of domestic currency. These price effects should induce a rise in exports and a fall in imports with consequential benefits to domestic activity. In turn, providing certain conditions hold (see below) the balance of trade and payments should improve thereby easing any potential constraint on domestic activity arising from the external account. Thirdly, in a deflationary environment such as that of the 1930s, devaluation should provide some relief from deflationary pressures; internal prices will fall less or rise relative to those abroad thereby imparting greater confidence to the business community.

Whether devaluation is effective along the above lines will depend upon several conditional factors. Preferably exchange depreciation should not occur abroad otherwise the comparative advantage will be reduced or even eliminated altogether, depending on the size of the devaluations elsewhere. Secondly, if the volume of exports is to be raised significantly the demand for exports must be relatively price elastic. Similarly, in order to produce a favourable movement in the balance of trade, the price elasticities of demand for both exports and imports must be favourable. To be effective the devaluing country must receive a constant or higher return for its exports and/or a reduction in payment for its imports expressed in foreign currency. Since however depreciation lowers export prices expressed in foreign currency the price elasticity of demand for exports must be equal to or greater than one in order to maintain or raise the level of foreign-exchange earnings. Conversely, since import prices in terms of foreign currency are left unchanged by devaluation (at least initially unless foreign exporters adjust their prices) then an elasticity of demand for imports of anything above zero will reduce the demand for imports and thereby lower the foreign currency bill. In other words, the balance of trade and balance of payments will tend to improve providing the sum of the elasticities for exports and imports is greater than unity, while if less than unity it will deteriorate. In the latter case an appreciation would be required to improve the balance. A positive response also assumes that supply elasticities are favourable and that the deficit in the balance of payments is not large. The larger the initial deficit the greater will need to be elasticities to eliminate it. Finally, it is possible for the balance of trade to improve even when the sum of the demand elasticities is less than unity, providing the supply elasticities of imports and exports are very low.

Even if these demanding conditions obtain, the size of the improvement may be offset by other factors. If the devaluing country is an important world trader then other countries may take exception to its action and retaliate, not necessarily by competitive devaluation but by increasing restrictions on imports through tariffs or quotas, thereby reducing the scope for the devaluing country to increase its exports. Alternatively, exporters may adjust their prices for exports and partly offset the price effects of devaluation. For example, domestic exporters may take advantage of the situation and raise their prices, especially if devaluation generates higher domestic inflation and cost pressures (wages), while foreign exporters may absorb part of the unfavourable currency effect by lowering their prices. There are, too, important income effects which must be taken into account. If devaluation raises exports relative to those of the rest of the world then incomes will presumably rise faster than elsewhere. Now if the devaluing country has a relatively high marginal propensity to import then a significant part of the increase in domestic incomes will spill over into imports which will partly offset the potential decline in imports through the currency price effect. Broadly speaking, such income changes will tend to dampen the balance of payments effect of exchange depreciation but they will not normally eliminate it altogether. Apart from these secondary income changes, the income elasticities of demand for imports and exports may be of considerable relevance to the consideration of trade movements at times when incomes are rising or falling rapidly at varying rates in different countries, as in the early 1930s, since they can offset or even eliminate the currency price effects.

Clearly then the results of devaluation will depend upon the interaction of many different factors, though as a general rule one would expect some competitive gain to accrue to the devaluing country. As far as the British devaluation is concerned, it is also important to bear in mind the unusual circumstances in which it occurred. It took place near the trough of a severe depression at a time when both export and import prices were falling rapidly, and when incomes abroad were falling faster than in this country. It was followed soon after by general tariff protection, and by increasing trade restrictions and exchange control in other countries. Such factors obviously complicate any assessment of the impact of the British devaluation.

Let us now turn to the facts of the situation. On *a priori* grounds one would expect some initial gain from devaluation given the fact that sterling is generally held to have been overvalued prior to 1931. However, Britain was by no means the only country to devalue. A number of countries, mainly debtor primary producers, had devalued before Britain, while in 1931 most British Empire countries and a number of others left the gold standard and pegged their curencies to the pound sterling. Then in the year following sterling's devaluation a further twelve countries relinquished the gold standard, while another wave of currency depreciation occurred when the dollar was devalued in 1933. Consequently, by the end of 1933 most major currencies and many minor ones had been devalued; the chief exceptions were those of France, Belgium, Italy, Holland, Switzerland, Poland and Czechoslovakia, which formed the hard core of the gold bloc group and which maintained their currency values until the mid 1930s. In addition, Germany and many central and south-eastern European countries practised rigorous exchange control. The magnitude of the depreciations varied considerably — some being larger than the British — but clearly Britain's initial advantage was steadily whittled away by similar action elsewhere.

This does not mean however that Britain failed to reap any benefit from her early move, though clearly the gains fell short of what might be expected from an initial reading of individual bilateral exchange-rate movements. When the gold standard was abandoned in September 1931 the pound was allowed to float[1] and at the end of the first week it had depreciated by nearly one quarter against the dollar; by the beginning of December it reached a low point of $3.40, or one third below its former parity. A similar descent was recorded against the French franc and other currencies which remained tied to gold. However, the bilateral rates of key currencies give a misleading impression of sterling's effective rate on a trade-weighted average of bilateral rates, since sterling's movement over a range of currencies was far from uniform. Thus in contrast to the initial sharp fall against the dollar and the franc, Britain's average effective rate fell by only 13 per cent between the third and final quarters of 1931. Subsequently the pound appreciated against the dollar and depreciated further against the franc; between 1932-34 it rose 40 per cent against the dollar and by the latter year it was above the former parity value, whereas in the same period it fell by 14 per cent against the franc. Redmond's effective rate, as measured against a basket of 28 currencies (1929-30 = 100) suggests an average depreciation of about 13 per cent in 1932, 9 per cent in 1933 and 4-4½ per cent until well into 1936, after which it rose above its base level.[2] On the other hand, as with the assessment of sterling's overvaluation in the period 1925-31, the extent of the adjustment in 1931 and the subsequent movement in sterling's effective rate depend very much on the price indices which are used to measure relative purchasing power. What seems certain however is that the pound's overvaluation was corrected in 1931, possibly leading to undervaluation in 1932, but thereafter the initial gain steadily evaporated as other countries devalued and the pound appreciated.[3]

Nevertheless, the effects of Britain's initial currency advantage in terms of exports can scarcely be described as dramatic. Depreciation certainly did not lead to an upsurge in exports though in the early years Britain did secure a relative trade gain. Exports recovered earlier than those of some major countries such as the

United States, Germany, France and other gold bloc and exchange-control countries, and through to 1935 Britain's exports outstripped the growth in world trade. Thereafter Britain's relative trade performance deteriorated so that over the complete cycle, 1929-37, she barely managed to maintain her trade share in manufactures. World trade in manufactures fell by 17.5 per cent between 1929-37, whereas the British decline was over 20 per cent, so that her trade share declined slightly from 23.6 per cent in 1929 to 22.4 per cent in 1937. Even so this was a better achievement than in the previous decade when Britain had one of the worst export records, whereas in the 1930s several countries turned in an inferior performance.

Only part of this relative improvement can be attributed directly to the price effects of devaluation, and most of the gains came in the first two or three years. The relatively early devaluation helped to check the fall in export volumes, in contrast to the experience in many other countries, with the result that in the years 1931-33, possibly into 1934, Britain was able to secure a slightly larger share of world trade even though total export volume remained decidedly sluggish. Thus the average level of exports in 1932-34 as against 1931 was considerably above the corresponding relative levels of the United States, Germany, Japan and the gold bloc countries. It is interesting to note, however, that the greatest export gains were made in countries with depreciated currencies as opposed to those still on gold where one would expect the potential gains to be larger. After devaluation the percentage of exports going to non-depreciated countries declined, it rose with those countries with less depreciation than Britain, while there was an even greater increase in share to those countries with larger devaluations than the UK.[4]

This would appear to indicate that the income effects of devaluation were more important than the price effects. Countries which left the gold standard and devalued their currencies were more prosperous or experienced earlier recovery than those still on gold who were forced to deflate to maintain their exchange rates, and were therefore in a stronger position to increase purchases from each other. The price effects of devaluation were probably relatively limited despite the fact that initially Britain was able to offer her exports on fairly favourable terms compared with 1931. This can be explained by the low elasticity of demand for British exports due to sharper declines of income in some of Britain's major markets, especially primary producing countries, to the extensive depreciation of other currencies, and to the continued deflation and increasing restrictions on trade in countries still on gold and in exchange-control countries. In other words, it was the revival of activity and incomes in countries that had left gold which helped to make the British devaluation effective, insofar as it can be considered to so be.

The importance of the indirect income effects should not be underrated. The improvement in the British trade balance between 1932-35 was associated in part with the early and relatively large improvement in the condition of some agricultural and primary producing countries, and exchange depreciation contributed to this improvement. It brought financial relief to many debtor countries and enabled them to pursue policies conducive to recovery. There were also other factors working in the same direction. Imperial preference fostered stronger trading links with the Dominions and the Colonies, while trade agreements with foreign countries sometimes helped to stimulate exports.

Thus while in no sense could it be argued that devaluation brought about an export-led recovery, it did help to stabilise the level of British exports and in the first few years Britain did have a competitive edge over many other countries. On the import side the visible impact is more discernible but unfortunately it is difficult to determine exactly how much of this was due to devaluation as opposed to tariff protection, both of which worked in the same direction. Devaluation had little impact of course on imports of food, raw materials and fuel, the demand for which was relatively inelastic. The main effect was on manufactured imports. These fell dramatically in 1932, by no less than 44 per cent, and remained fairly depressed for much of the decade so that by 1937 they were still 17.6 per cent lower than in 1929. This may be contrasted with a 66 per cent rise in manufactured imports during the 1920s. Given the growth of output between cyclical peaks, 1929-37, the volume of manufactured imports was abnormally depressed in the 1930s. Devaluation of sterling probably played a relatively small role in the decline since the large fall which took place in 1932 was mainly accounted for by the imposition of the general tariff. There is moreover no evidence of a strong decline in imports in the months immediately following devaluation — in fact they were rising late in 1931, though no doubt partly in response to the imminent threat of protection. After 1933 manufactured imports began to rise again in line with the growth of total output.

Sorting out the effects of depreciation on the balance of payments is an even more intricate task. As noted earlier, the current-account balance deteriorated sharply between 1928-31, from a substantial surplus to a record deficit of £114 million. This resulted from a deterioration of the merchandise trade balance and a sharp fall in invisible income from abroad. Despite an improvement in the terms of trade (which *ceteris paribus* would have improved the balance) import values fell very much less than export values with a consequent widening of the deficit. The modest fall in British incomes compared with those abroad coupled with the low elasticity of demand for imports meant that imports were maintained at a relatively high level, whereas the sharper fall in world incomes and the high income elasticity of demand of foreign countries for Britain's exports caused a substantial decline in this country's exports. Thus income changes were again more powerful than the price effects.

Depreciation had little effect on the trade balance one way or another in 1931. Since it was only operative in the last quarter of that year it affected the average exchange rate for the year as a whole by only 7 per cent, whereas export unit values fell by one sixth and import unit values by one quarter (in dollar relatives), thus giving rise to an 11 per cent improvement in the terms of trade.[5]

During 1932 and 1933 the balance of payments improved considerably and the current-account deficit was almost eliminated. Most of this improvement derived from the trade account, the adverse balance declining from £322 million in 1931 to £192 million in 1933. Since the bulk of this improvement occurred in 1932 it cannot be attributed to the terms of trade, which deteriorated slightly in that year. Nor does it seem likely that depreciation contributed significantly to restoring equilibrium in the external account. In the short run the demand for many imports was relatively inelastic with respect to price changes, while the price elasticity of demand abroad for British exports was fairly low because of the income effects. In fact recent estimates suggest that the sum of the price elasticities was slightly below unity (0.95), which implies a short-term worsening of the trade balance.[6] Thus, devaluation may, as argued above, have stabilised the level of exports but it was from the import side that the main relief came. Imports were cut back sharply in both value and volume in 1932 and there was a further decline in values in 1933. The price effects of devaluation had only a marginal impact in the squeeze on imports. The factors mainly responsible were the tariff, the continued decline in import prices and, to a lesser extent, the reaction to the anticipated purchases of 1931.

Thus devaluation made relatively little contribution to the improvement in the external account through to 1933. Thereafter, apart from 1935 when a small surplus was recorded, the current-account balance deteriorated modestly through to 1938. For the period as a whole the average annual deficit was £37 million compared with an annual surplus of nearly £22 million between 1925-31. Despite an improvement in the invisible balance, the trade balance deteriorated steadily after 1933 as imports responded to the pace of domestic activity, the terms of trade deteriorated and export recovery faltered in the mid-1930s. In addition, the steady appreciation of the effective exchange rate did not help matters either, despite attempts by the Exchange Equalisation Account to keep the rate down. The upward pressure on the exchange rate was enhanced by the tariff and by the continued control on overseas lending which had been imposed at the time of the suspension of the gold standard. Thus insofar as the Treasury wished to see a lower exchange rate it could have relaxed the control on foreign lending, which was the counterpart to a strong exchange rate for balance of payments equilibrium.

Overall, therefore, depreciation of the currency only made a modest contribution to Britain's recovery effort. The direct benefits were both limited and temporary. It did not initiate any export-based recovery and the improvement in the external balance can be attributed largely to other factors. It was not until 1934 that exports really began to show signs of recovery and by that time much of the initial benefits of devaluation had been neutralised by counteraction abroad. On the import side the depreciation was swamped by the effects of the tariff. On the other hand, devaluation need not be written off completely. For a while Britain secured a comparative advantage in world markets and this helped to stabilise the level of exports, while the spread of depreciation abroad contributed to a rise in activity in other countries which eventually reacted favourably on Britain's exports. Secondly, exchange depreciation helped to modify or check the deflationary tendencies in Britain, at first by a relative rise in prices, that is a smaller decline than elsewhere, and later by an absolute rise. This increase made for a more favourable cost-price structure in industry and thereby increased business confidence. This, it has been suggested, was more significant than any trade gains. 'Exchange depreciation', according to Harris, 'seems to help more by improving the price-cost structure at home, thus contributing to a more satisfactory business situation, than by capturing foreign trade at the expense of rivals.'[7] Moreover, devaluation did not result in a deterioration in the terms of trade, which improved by a quarter between 1929-33. This was a major factor in the high floor to incomes and consumption during the depression which helped to pave the way for an early recovery (see Chapter 6). Finally, and possibly of greatest importance, the break with gold provided the authorities with room for manoeuvre on the domestic policy front. Since the monetary authorities were no longer constrained by the necessity of defending a fixed parity it was possible to ease the monetary stance once they had overcome the fear that devaluation would lead to uncontrollable financial pressure and inflation. Capital controls, tariffs and the strengthening of the exchange in 1932-33 eased the transition process to a regime of cheap money, the effects of which are discussed in the next chapter. Whether the removal of the parity constraint would also have allowed a more liberal fiscal policy, had the authorities been so inclined, is somewhat more debatable given the critical importance of restoring financial confidence after the crisis of 1931 and the possibility of a balance of payments constraint arising had a more expansionary fiscal stance been adopted (see Chapter 4).

## General Tariff Protection

Until the early 1930s Britain was still very much a free-trade country with some 83 per cent of her imports being allowed in duty-free. The remaining imports were subject to varying degrees of duty, most of which represented revenue duties which had been imposed during the nineteenth century. Protective duties affected no more than 2-3 per cent of imports, and many of these duties had been levied during the war or shortly afterwards on key or strategic products. The list included motor cars, synthetic dyestuffs, scientific instruments, wireless valves, magnetos and a number of other products. Most of these duties remained in force after the adoption of general tariff protection in 1932.

The abandonment of free trade was effected in a number of stages. Following the departure from gold in September 1931 and the election in the following month which returned a large Conservative majority in the National Government in favour of protection, an emergency measure, the Abnormal Importations (Customs Duties) Bill was rushed through Parliament in an effort to check the surge in imports anticipated while more permanent legislation was being debated. This gave the Board of Trade temporary power to impose duties of up to 100 per cent *ad valorem* on imports (mainly manufactured or semi-manufactured products) which were thought to be entering the country in abnormal quantities. Duties of 50 per cent were imposed immediately on a wide range of manufactured goods. Horticultural products were also protected by special legislation which came into force before the end of the year. Early in the following year the basis of a permanent policy of protection was established by the Import Duties Act. The emergency duties were replaced by a general 10 per cent *ad valorem* tariff on all imports, except those on a free list which included most raw materials and foodstuffs, and a new tariff-making body, the Import Duties Advisory Committee, was set up to consider applications for revisions in the tariff structure. Following its first report in April 1932, duties on manufactured goods were raised to 20 per cent *ad valorem* across the board, with higher imposts recommended for some luxury goods, chemicals, and certain industrial raw materials and semi-manufactures. In the following year alterations were made to specific rates, the most important being the duty of 33⅓ per cent on steel imports (temporarily raised to 50 per cent in 1935) in an effort to induce the industry to reorganise.

The new tariff arrangements did not apply to Empire products, but they formed the basis for imperial preference agreements negotiated at the Imperial Economic Conference held at Ottawa in the summer of 1932. At this conference the principle of trade discrimination in favour of Empire products was confirmed. Britain gained minor tariff concessions and increased preference for British products in Dominion markets in return for which she guaranteed free entry for most Dominion products as well as increased margins of preference by imposing new or additional duties on certain goods imported from foreign countries which competed with Empire exports. In 1933 the Imperial preference scheme was extended to the Crown Colonies. Finally, a system of quota restrictions on foreign food imports together with certain quantitative limitations on Empire imports were imposed to safeguard the British farmer.

The overall effect of the new legislation can be summarised as follows. Most Empire and Colonial imports were admitted duty-free or at nominal rates of duty, whereas the greater part of foreign imports was subject to varying degrees of tariff protection. After the Ottawa negotiations, the percentage of imports which came in free of duty was about one quarter (though some were restricted by other means) as against 83 per cent in 1929-30. Tariffs of between 10-20 per cent were levied on just over one half of foreign imports, nearly 8 per cent paid duties of over 20 per cent, while the remaining 17 per cent were still covered by the old revenue and safeguarding duties. Tariff levels tended to be more severe on products from European countries. The highest rates of duty were levied on manufactures and semi-manufactured articles and the lowest on food and raw materials, a considerable proportion of which were on the free list. Though there were no major changes in the general tariff structure after 1933, the levels of tariffs on various categories of goods were revised from time to time following recommendations of the Import Duties Advisory Committee. On balance the trend was in an upward direction and by the end of the decade the number of articles subject to duty and the average rate of duty levied were higher than in 1932. In addition, a number of the earlier *ad valorem* tariffs were replaced by more onerous specific duties. Finally, the government used the new tariff structure as a bargaining counter to negotiate a series of bilateral trade deals culminating in the Anglo-American Agreement of 1938.

Before analysing the effects of the new tariff structure, it might be instructive to pause for a moment to consider the motives behind the shift to protection. In a world rapidly taking refuge in all sorts of restrictions on trade and with a newly elected Parliament dominated by a strong protectionist lobby, it is perhaps not surprising that Britain made the radical departure in trade policy when she did. But the objectives and benefits of tariff protection remained confused and contradictory in the minds of contemporary statesmen and economists. Chamberlain, in his speech introducing the general tariff legislation to the House of Commons early in 1932, made vast and unsubstantiated claims for the new policy. He defended it on the grounds that it would produce untold benefits; it would, he said, raise revenue, improve employment, hold down imports, improve exports, strengthen the links with Imperial countries and even avert a rise in the cost of living following devaluation.[8] While some of these projected benefits were mutually contradictory, it is clear that the recovery properties of tariff protection feature prominently. Similarly, in the year immediately prior to the abandonment of gold, politicians and economists could be found debating the case for a tariff as a means of creating employment, raising the price level and protecting the external account without having to relinquish the gold standard.

Once the decision had been taken to break with the sterling party, there appeared to be less justification for a tariff on employment and balance of payments grounds assuming that a floating exchange would take care of the external balance and thereby allow a domestic stimulus to be applied. For this reason Eichengreen questions the assumption that the tariff was implemented as an employment device. Rather he interprets it as a means to strengthen the trade balance and prevent the exchange rate from depreciating. This would tend to square with the thinking of Ministers and Treasury officials at the time who remained unconvinced that a floating rate would solve Britain's external problems, and who believed, what was worse, that it could lead to a spiral of inflation and currency depreciation together with budgetary deficits. Thus the restoration of external balance and the stablisation of the exchange rate were, at the time of the tariff debate, deemed to be more important as a means of restoring confidence than the employment and recovery potential of a tariff. Indeed, it was generally recognised that insofar as a tariff was likely to strengthen the sterling exchange it would be counter-productive from an employment point of view, which is one reason why the Treasury soon became anxious to keep the pound down. Eichengreen concludes therefore that, contrary to many studies which see the adoption of protection as a misguided employment policy, the evidence suggests that it was initially adopted because of the authorities' distrust of the likely repercussions from a floating exchange rate. Other factors which can be identified as affecting the shape of the tariff structure were the strong support for Imperial preference and the need to rationalise the staple industries.[9]

The task of analysing the impact of the tariff is therefore a complex one given the conflicting views as to the nature of its objectives. Moreover, the ramifications of protection were inevitably varied and often of an indirect nature. It is also difficult to sort out the effects of protection from those caused by other factors such as devaluation, cheap money, improved terms of trade and the natural forces of recovery, all of which occurred at a similar point in time. However, it is important to attempt an assessment of the broad impact of the tariff on Britain's trade, industry and employment and the way in which it affected the structure and pattern of trade.

Generally speaking one would anticipate that the introduction of tariff protection would lead to a fall in imports and an improved trade balance. But the direction and magnitude of the changes is conditioned by a number of factors. Much will depend upon the level of the tariff and its coverage, its timing, the elasticity of demand for imports, the terms of trade and the behaviour of exports. A low tariff with a wide coverage, for example, may be too weak to offer an effective check to imports, while even a moderate or severe tariff may have a limited impact if import demand elasticities are very low. Alternatively, if foreign exporters absorb much of the cost of the tariff, by reducing their selling prices, imports in the receiving country may not respond favourably. Assuming however that imports are reduced by tariff protection, the balance of trade may not necessarily improve because of adverse movements on the export side. Tariff protection will tend to make the domestic market more attractive thereby inducing manufacturers to switch resources away from export production, a trend more likely to occur during boom conditions than at times of under-utilised resources. Secondly, the demand for British exports may decline following relatiatory tariff action abroad. Thirdly, tariffs will increase the cost of imported materials (assuming these are subject to duty) thus raising the cost of producing exports and lowering competitiveness. Conceivably therefore, some of the gains derived from tariffs through reduced imports may be offset by adverse repercussions on the export side.

The possible permutations are almost unlimited and the acid test is again recourse to the facts of the particular situation. The short-term effects were undoubtedly favourable partly because of the fortuitous timing of the tariff. Imports were running at a high level in the latter half of 1931, no doubt in anticipation of protection, and the immediate response to the new duties was a sharp fall in imports. The total volume of imports fell by 13 per cent in 1932 and in the following year they stayed flat. There was however considerable variation among different categories of imports. Food imports declined only moderately and then against a peak level in 1931,[10] while imports of fuel and raw materials actually rose through 1931-33. Such trends might have been anticipated given the lighter incidence of restrictions on these goods and their low demand elasticities. It was on manufactured products that the tariff bore heavily. Retained imports fell by nearly 44 per cent and much of this fall can be attributed to protection rather than to changes in domestic incomes or the price effects of devaluation. The decline in domestic incomes was very modest in 1931-32 and was more than offset by a sharp rise in the following year, while the price of imported manufactures in sterling terms (excluding the tariff) actually fell.[11] The drop in the total volume of imports was reinforced by the declining prices so that in value terms imports fell by more than volume, from £786 million to £619 million between 1931-33, a decline of over 21 per cent.

The indirect short-term effects of the tariff on exports are not easily discernible from the trade data. It is probable that any indirect adverse repercussions were compensated by the initial benefits arising from depreciation and the beginnings of recovery. In any case, exports had reached their nadir in 1931 before general protection was imposed, though tariff levels generally were rising abroad. The volume of exports remained stable in 1932 and 1933, though there was a modest fall in values between 1931-32 as a result of price changes. Thus the significant improvement in the trade balance over these years was largely due to changes on the import side.

Tariff protection could not hold down the flow of imports indefinitely. By 1934 they were rising strongly and they continued to do so until 1937, when they exceeded the previous cyclical peak of 1929 by 8.1 per cent. This increase was largely to be expected in the recovery stage of the cycle. The question is whether protection damped down the volume of imports to a lower level than they would otherwise have been. During the recovery phase 1932-37 the total volume of imports rose by 23 per cent, which is very similar to the rise in GDP though rather less than the growth of industrial production. Between cyclical peaks (1929-37), on the other hand, import growth was only one half that of GDP and less than a third that of industrial production. Since in both periods imports of food, fuel and materials rose significantly, it was in manufactured imports that the main setback occurred. While the latter showed a sharp recovery between 1932-37, which was well in excess of either GDP or export growth, it should be stressed that they started off from a very low base and failed to regain their previous peak. Through 1929-37 they were down by 17.6 per cent whereas GDP rose by about the same amount and industrial production increased by 30 per cent. Thus it would appear that protection caused a temporary break in the trend of imported manufactures which was resumed as recovery got under way. In part this may be attributed to the relatively low level of British tariffs which were often found to be insufficient to offset the competitive advantages of overseas suppliers. In a study of 25 commodities for 1937, MacDougall found that our tariffs offset the American competitive advantage in the British market in only two commodities, paper and glass containers. By contrast, in those goods in which Britain had a price advantage in the US market, the level of tariffs there was generally high enough to offset it.[12]

The British tariffs were generally too low to act as a permanent deterrent to imports. In any case, the effects were partly mitigated by price cutting and dumping in the home market by foreign firms with excess capacity. Conversely, British exporters faced higher tariffs abroad and they were in a weaker position competitively. The tariff also had the effect of inducing manufacturers to pay more attention to the domestic market in the 1930s. Thus exports rose only modestly in this period and continued to remain well below the 1929 level. It is true that export volumes rose slightly faster than imports in the recovery phase, but then they had fallen far more heavily in the slump. But the rise in export volumes was partly offset by adverse price trends, especially the rapid rise in import prices in the later 1930s so that, in terms of value, imports rose faster than exports with a consequent deterioration in the trade balance.

One important structural change in the interwar years was the decline in the importance of trade in terms of output, and the concomitant growth in the relative importance of the domestic market. The climax of this development came during the 1930s and some writers have associated it with tariff protection. In fact, however, the dwindling importance of international trade predates this period. Exports and imports as a percentage of national income peaked in 1913 after which the share declined. By 1931 imports as a proportion of national income were 20.8 per cent as against 31.1 per cent in 1913, while the respective shares for exports (including re-exports) were 12.1 and 28 per cent. Following the introduction of protection the shares did not change very much. By 1938 the import-income ratio was 17.6 per cent and that for exports 11.1 per cent. In fact after 1933 the shares remained remarkably stable and it is clear that the trend towards greater self-sufficiency had made its major advance before protection was introduced.

With the decline in international trade and the contraction of Britain's staple export industries it was more or less inevitable, irrespective of tariffs, that recovery would be based primarily on the home market. This does not mean of course that the tariff had no effect on the recovery process. The initial cutback in imports of manufactures and the strengthening of prices no doubt acted as a boost to business confidence and may have helped to stimulate output and investment in some protected industries. Some writers have argued that the tariff was an important stimulus to output, employment and prices (via the terms of trade) but one should be cautious about accepting the extravagant claims of the architect of IDAC that it 'proved on balance of significant and essential value in aiding Britain's recovery, not only by giving an initial stimulus to the rise from the slough, but in the longer term also.'[13] Whether this bold assertion is accurate or not depends very much upon the extent to which newly protected industries were responsible for recovery and whether their rate of development was in any way associated with tariff protection.

As far as the first point is concerned, it is worth noting that the initial upturn from the depression was far from being based on newly protected industries. Indeed, some of the pace-setters in the early phase of recovery were sectors largely unconnected with foreign trade, such as construction, service industries, and motor manufacturing which had been protected prior to 1932. Secondly, many of the industries which first received protection in 1932 were the old staple trades and they contributed very little to the beginnings of recovery. At best protection probably checked output from falling further in these sectors but in the longer term it could have had adverse consequences by sheltering relatively inefficient industries and delaying much needed reorganisation and capacity elimination. Thirdly, the relationship between output, employment and imports in the newly protected industries does not lend a great deal of support to the hypothesis that tariffs played a crucial role in industrial recovery. In a survey of twenty industries Richardson found that the rate of employment growth between 1930-35 was very modest and no greater than in manufacturing industry generally. Furthermore, there was no clear-cut association between movements in imports and the growth of employment and output of these industries. In some cases where imports fell sharply, as in machinery, cotton goods, wool manufactures, clothing, leather goods and hats and caps, rather modest rises in output were achieved, whereas certain rapidly growing industries such as chemicals, glassware, pig-iron and non-ferrous metals, experienced fairly small declines in imports. On the other hand, in toys and steel products, where the fall in imports was above average, expansion was rapid.[14]

The debate on the recovery properties of the tariff still goes on. Recent work by Forrest Capie takes a particularly gloomy view of the outcome. He maintains that its role in stimulating the upturn was insignificant with little evidence generally that protection did anything to promote economic growth and welfare. Two of the more important sectors in the early phase of recovery, iron and steel and building, had low or negative effective rates of protection, while actual resource flows did not always correspond with the degree of protection. Industries with high effective rates included glass-making, cotton, woollen goods, clothing and motor vehicles, whereas those with low rates comprised some of the new industries, iron and steel, shipbuilding and certain types of glassware. Across the board analysis of effective rates would appear to indicate that protection slowed down the structural changes required by protecting the old staples such as wool and cotton, while giving less protection to the newer industries.[15] On the other hand, Foreman-Peck, in a critique of Capie's earlier work, sees a more positive advantage with a modest contribution being made to output and employment, of the order of two percentage points of GDP.[16]

On balance, therefore, one might conclude that tariff protection was not of vital significance to the recovery effort. Moreover, in the later 1930s, when imports of manufactures were rising sharply, it must have exercised even less influence. But one should not conclude that there were no beneficial consequences arising from tariff protection. The fact that most newly protected industries experienced some respite from imports in the early 1930s certainly helped to improve business expectations and optimism about the future. The relative lack of association between changes in imports and rates of growth in protected industries, though suggestive, is not conclusive proof that tariffs were completely ineffective unless a *ceteris paribus* condition holds true. If furthermore one were to speculate on what might have happened in the absence of protection, then the case for tariffs is somewhat strengthened. With excess capacity in many competing countries, Britain, as the only free market, might well have been flooded with cheap (artificially priced) imported manufactures on an even greater scale than in the 1920s, when imports more than doubled from the trough of 1921. This would have depressed business confidence, prolonged the depression in Britain, especially in some of the old staple industries, and weakened the forces of recovery. The tariff, at the levels set, could not provide complete protection from imports, nor did it rank as a critical element in the recovery, but it did offer some much-needed relief to certain industries, especially those which had been hit severely by imports before 1932. Thus despite its apparently modest contribution to recovery, in the economic conditions prevailing at the time, free trade by one country alone was untenable.[17] Maybe it is time the critics of the tariff explored more closely the counterfactual world.

## Trading Agreements and Imperial Preference

The use of the tariff weapon to stimulate domestic activity (assuming that imports are responsible for under-utilised capacity) may well turn out to be self-defeating if countries abroad retaliate in a similar manner thereby making it more difficult for the country first imposing trade restrictions to export its own goods. This line of argument is not particularly applicable to the British case at the time in question. In the first place the pattern of Britain's development after the war made it very unlikely that exports would be the leader out of depression. Secondly, exports had more or less reached their trough before general protection was introduced, though the tariff may have had a marginal drag effect on the subsequent recovery rate of exports. Thirdly, tariffs were fairly widespread abroad by the early 1930s so that Britain's move to protection did not spark off a new tariff scramble. It is true that many countries raised their tariff levels and imposed other restrictions on trade, but this was more a response to trade depression than a reaction to Britain's departure from free trade. In fact, far from arguing that Britain's export prospects were harmed indirectly by the response elsewhere to British protection, it can be claimed that given the prevalence of tariffs and other trade restrictions the world over, Britain required a tariff to use as a bargaining counter in trade negotiations. Without such a weapon this country would have been powerless to compete in a world riddled with trade restrictions. In contrast to the position before the war, and to a lesser extent in the 1920s, the international trade system of the 1930s represented 'a series of exclusive bargains between governments on the basis of strict reciprocity' and the tariff became a vital instrument for exacting concessions.[18]

Thus in common with many other countries, the British government attempted to improve the position of British traders by concluding a series of bilateral trade and clearing agreements with her main trading partners. The control of trade and payments along bilateral lines was never as extensive as in some European countries, notably Germany, but it certainly represented a radical departure from previous practice. By 1938 nearly one half of Britain's total trade with foreign (non-Empire) countries was conducted through bilateral trade, payments and clearing agreements, while the Ottawa negotiations established the system of Imperial preference with the tariff being used as a bargaining counter. In addition, specific tariff increases were sometimes threatened to extract concessions from foreign producers. The steel tariff, for example, was raised to 50 per cent in 1935 in an effort to force the European Steel Cartel to curtail its exports to this country. As a rule Britain adhered throughout to the principle of the most-favoured-nation clause which meant that she refrained from discriminating in her own tariff and quota policies between different countries. The main aim of bilateral bargaining from the British point of view was to induce foreign countries to discriminate in her favour in an effort to stimulate exports and improve debt collection.

A study carried out by the National Institute during the second world war claimed this to be one of the most important and successful features of government policy during the 1930s.[19] It is difficult to see how this conclusion was derived since the measurable effects of the agreements on Britain's trade can have been but slight. The series of bilateral trade deals negotiated with foreign countries, the most important of which were with the Scandinavian countries, certain Eastern European countries and South America, brought remarkably few reductions in tariffs, and the concessions on either side were very small. The major exceptions to this rule were the agreements made with France in 1934 and the United States in 1938. The latter — the Anglo-American trade agreement — was notable for the fact that it led to the first appreciable reduction of the British tariff since 1932 and involved a positive reversal of the policy of Imperial preference. This latter issue proved one of the chief stumbling blocks to Britain granting significant concessions to foreign countries since to have done so would have broken the pledge to the Dominions to maintain a minimum margin of preference. In general the most that Britain could do was to promise that duties would not be increased.

As far as trade between contracting parties was concerned, this tended to increase and with some of the trade agreement countries, notably the Scandinavian and Argentinian, British exports made headway, though in other cases, notably with Norway and Finland, they lost ground. One of the most successful ventures was that with Denmark where the Import Duties Act of 1932 was used to force Denmark into making extensive and unreciprocated concessions which gave British exports a share of the Danish market quite out of keeping with their relative competitive strength. In value terms Britain's exports to Denmark rose from £8.7 million in 1931 (1929 = $10.7 million) to £16.9 million in 1937, whereas imports from Denmark fell from £46.7 to £36.6 million (1929 = £56.2 million) during the same period.[20] Similarly, in one or two other cases in which Britain was able to exert commercial leverage she succeeded in exacting highly favourable concessions. This was certainly true in the case of Argentina, a country very dependent on the British market. In a deal negotiated in 1933 Argentina agreed to allocate most of the sterling proceeds of her exports to the UK for purposes of debt repayment, both public and private, in return for Britain promising to maintain her meat imports at the Ottawa level. Two years later Britain also secured a satisfactory undertaking with Germany on export proceeds and commercial debt collection.[21]

These successes were, however, the exception rather than the rule. Overall Britain probably gained slightly in terms of trade and debt collection through the series of agreements. On balance, exports to trade agreement countries rose faster than imports from them, though Benham argued that British exports would have increased more had the agreements never been made.[22] At best the total effects were very small, and some of the gains were in fact offset by adverse indirect repercussions from the bilateral arrangements in the form of increased competition in third markets from countries excluded from the agreements. The most notable instance in this respect is that of British coal in the Scandinavian market, the demand for which was raised by agreements concluded in 1933 and 1934. As a result Scandinavian countries reduced their purchases of coal from Germany and Poland, wherupon the latter two countries intensified their competition in Mediterranean and other countries not covered by agreements, to the detriment of British exporters. Thus the gains secured by Scottish and North-eastern coal producers in Scandinavia were offset by the losses suffered by producers in South Wales exporting to markets where German and Polish competition had increased.[23]

Finally, we turn to the Ottawa agreements and Imperial preference. Since the Empire countries were largely exempt from the original tariff of 1932, Britain was able to extend preference to the Dominions only by raising restrictions (tariffs and in some cases quantitative restrictions) against foreign imports, which included a number of new or additional 'Ottawa duties' on certain foreign foodstuffs the most important of which were wheat, maize, butter, cheese and other dairy products. In return Britain secured some minor tariff concessions and increased preferences for British goods in Imperial markets, though generally this took the form of an increase of their tariffs on foreign goods rather than a reduction of those on British goods. Thus, although this meant a greater degree of liberalisation of intra-Empire trade, it did lead to an increase in restrictions against the rest of the world.

How far Imperial preference fostered the economic links of the Empire is difficult to assess since prior to 1932 the trend had been in the direction of greater Imperial unity. Britain had of course always had strong links with Imperial countries, and given this connection and the difficult trading conditions in many foreign countries — especially European — during the 1930s, it was only to be expected that Imperial connections would flourish. What can be said fairly certainly is that the Dominions appear to have been the main beneficiaries from the Ottawa arrangements. Between 1930 and 1938 British exports to the Empire rose from 43.5 to 49.9 per cent of total British exports, whereas over the same period imports from the Empire increased from 29.1 to 40.4 per cent of total UK imports. On the other hand, while Britain increased the proportion of her exports going to Imperial countries, the Dominions did not increase their share of imports from this country. Only Australia in fact recorded a significant increase in her share of imports from Britain.[24] In other words, while the UK was becoming steadily more dependent on the Empire as a market, the Dominions not only found a relatively secure market for their primary products at a time when market conditions were difficult, but they also were able to reduce their dependence on Britain as a source of supply. Moreoever, from Britain's point of view there were further disadvantages from the closer trading arrangements. As a consequence of Imperial preference British imports from non-Empire countries were reduced, thus weakening the ability or willingness of foreign countries to purchase in the British market. In addition, the increase in sterling made available to Imperial countries as a result of increased British imports was used to amortise debts or repurchase British-held investments rather than to buy British goods. The Imperial preference system also hampered Britain in bilateral negotiations with foreign countries and aroused the anger of both the British farmers and manufacturers, the one because many Empire primary products were admitted duty-free, the other because the Dominions continued to maintain fairly high protection on British goods. Finally, from a long-term point of view, it is debatable whether British exporters should have been paying greater attention to what could be regarded as downstream markets. The growth potential of the richer Dominion countries was limited because of their relatively small market size, while the markets of the more populous parts of the Empire and Crown Colonies were not the most rewarding from an income and technology point of view. Perhaps therefore the glitter of the Imperial connection was not quite as bright as it was made out to be at the time.

## Notes

1. The float ceased to be clean after the middle of 1932 following the establishment of the Exchange Equalisation Account, the purpose of which was to smooth fluctuations in the exchange rate by market intervention and to keep the pound from appreciating.
2. J. Redmond, 'An Indicator of the Effective Exchange Rate of the Pound in the Nineteen-Thirties', *Economic History Review*, 33 (1980), pp. 88-90; A. Cairncross and B. Eichengreen, *Sterling in Decline* (1983), p. 86.
3. J. Redmond, 'The Sterling Overvaluation in 1925: A Multilateral Approach', *Economic History Review*, 37 (1984), p. 531. Dimsdale's measure of competitiveness based on relative real exchange rates shows a rather larger gain, some 18 per cent until 1936 after which it declined to 8 per cent in 1938. N. H. Dimsdale, 'British Monetary Policy and the Exchange Rate, 1920-1938', *Oxford Economic Papers, Supplement*, 33 (1981), p. 333.
4. S. E. Harris, *Exchange Depreciation* (1936), pp. 121, 130, 132, 148, 149.
5. C. P. Kindleberger, *The Terms of Trade* (1956), p. 104.
6. T. J. Thomas, *Aspects of U.K. Macroeconomic Policy during the Interwar Period: A Study in Econometric History*, University of Cambridge, Ph.D. (1975), pp. 228-9.

7. Harris, *op. cit.*, p. 179.
8. F. Capie, *Depression and Protection: Britain between the Wars* (1983), p. 96.
9. B.J. Eichengreen, *Sterling and the Tariff, 1929-32*. Princeton Studies in International Finance, 48 (1981), pp. 1-2, 28-32.
10. This high level may have been due to the fear of restrictions. With 1929 as a base, food imports rose through to 1931 and then fell slightly in 1932-33.
11. M.Fg. Scott, *A Study of United Kingdom Imports* (1962), p. 169.
12. G.D.A. MacDougall, 'British and American Exports: A Study Suggested by the Theory of Comparative Costs, Part I', *Economic Journal* 61 (1951), pp. 699, 704-5.
13. H. Hutchinson, *Tariff-making and Industrial Reconstruction* (1965), p. 165.
14. H.W. Richardson, *Economic Recovery in Britain, 1932-9* (1967), pp. 247-51, 257.
15. F. Capie, 'The British Tariff and Industrial Protection in the 1930s', *Economic History Review*, 31 (1978) and *Depression and Protectionism* (1983), pp. 105-6, 123, 140, 143.
16. J. Foreman-Peck, 'The British Tariff and Industrial Protection in the 1930s: An Alternative Model', *Economic History Review*, 34 (1981), and the reply by Capie in the same issue. An earlier study by Foreman-Peck concluded that protection of motor manufacturing provided net benefits to the industry which outweighed any consumer loss. Motor manufacturing was of course protected long before the adoption of general tariff protection. J. Foreman-Peck, 'Tariff Protection and Economies of Scale: The British Motor Manufacturing Industry before 1939', *Oxford Economic Papers*, 31, (1979).
17. Cf. S. Pollard, *The Development of the British Economy, 1914-1980* (1983), p. 124.
18. R.C. Snyder, 'Commercial Policy as Reflected in Treaties from 1931 to 1939', *American Economic Review*, 30 (1940), p. 802.
19. National Institute of Economic and Social Research, *Trade Regulations and Commercial Policy in the United Kingdom* (1943), p. 12.
20. T.J.T. Rooth, 'Limits of Leverage: The Anglo-Danish Trade Agreements of 1933', *Economic History Review*, 27 (1984), pp. 212, 224.
21. H.W. Arndt, *The Economic Lessons of the Nineteen Thirties* (1944), pp. 116-17.
22. F. Benham, *Great Britain under Protection* (1941), p. 146.
23. Total coal exports actually declined between 1932-36.
24. D.L. Glickmann, 'The British Imperial Preference System', *Quarterly Journal of Economics* 61 (1947), p. 451; I.M. Drummond, *British Economic Policy and the Empire, 1919-1939* (1972), pp. 101-4.

# 4 Macroeconomic Policy in the 1930s

We now turn to a consideration of macroeconomic policy and its role in the recovery of the 1930s. The devaluation of sterling, the introduction of general tariff protection, and the control of foreign lending eased the constraints on the external side and provided the authorities with an opportunity to relax their stance on monetary and fiscal policy. A change in policy direction was delayed initially however, for fear of inflation and exchange depreciation and the need to restore financial stability. In the event the shift in policy occurred on only one front, in the emergence of cheap money. Fiscal policy, for various reasons, was not pressed into service to aid recovery.

## Cheap Money Policy

Though the devaluation of sterling lifted some of the pressure from the external account and loosened the dependence of British interest rates on those abroad, it was not followed immediately by a relaxation of monetary policy. In fact, during the financial crisis in the summer of 1931 the Bank rate had been raised from 2½ to 4½ per cent and then to 6 per cent in September at the time the gold standard was abandoned. Fears that exchange depreciation and unbalanced budgets would lead to inflationary pressures and a panic withdrawal of capital prompted the authorities to take firm action in an effort to restore financial confidence. Once conditions stabilised and funds began to flow back into London as confidence in sterling revived, it was possible to bring down interest rates and expand the money supply. In the first half of 1932 Bank rate was lowered progressively until it was down to 2 per cent by the end of June when the great War Loan Conversion operation was announced.

If at first the authorities were reluctant to introduce cheap money they soon became convinced that it was the right line of action to take. During the previous decade considerable criticism had been made about the restrictive effects of high interest rates, and the subsequent downswing in activity reinforced these criticisms. The absence of an active fiscal policy added weight to the view that monetary policy would have to be the main weapon for stimulating revival from the recession, a view reinforced by the deflationary stance of budgetary policy in 1931-32 in an effort to restore financial confidence. Reporting in July 1931 the Macmillan Committee urged that cheap money was essential to recovery[1] and within a year this had been adopted as offical policy. But the authorities arguably had more immediate aims in mind than recovery when considering the shift in policy stance. Cheap money was seen as a way of keeping the exchange rate down once it had stabilised, by deterring the influx of hot money.[2] Second, and more important, it provided a convenient way of trimming public expenditure and solving the budgetary problem by lowering the cost of servicing the national debt. The almost indecent haste with which the authorities launched the Conversion Issue is indicative of their eagerness to reduce the debt burden. At the end of June 1932, when Bank rate was finally lowered to 2 per cent, the Chancellor announced the conversion of £2,085 million of 5 per cent War Loan, representing some 27 per cent of the national debt, to a 3½ per cent loan. Through strong patriotic appeals to holders of the debt it proved an outstanding success, all but 8 per cent of it being taken up.[3] Further funding operations at even lower rates were carried out between 1934-35 and altogether over £100 million was lopped off the cost of debt servicing. As a result there was a substantial fall in the cost of debt service in terms of national income, from 8.3 per cent in 1932 to 4.6 per cent by 1935.

Whatever the reasons for the shift in monetary stance in the early 1930s, it is clear that the outcome was a much more liberal and stable monetary regime than that which had prevailed in the previous decade. Bank rate was held at 2 per cent throughout the period and the general structure of interest rates adjusted accordingly. Short-term Treasury Bill rates fell to one half of one per cent or less, while long-term yields (2½ per cent consols and industrial debentures) declined by over one third, reaching a low point in the mid-1930s after which they rose modestly. Real interest rates also fell significantly in contrast to the rising trend between 1925-31. The decline in fixed interest yields prompted a recovery in share prices from the latter half of 1932 and by the middle of the decade the index of industrial shares had doubled compared with the low point in the trough. At the same time there was a steady and almost uninterrupted expansion in the monetary base which contrasted sharply with the stagnation in the money supply through to 1931.[4]

Cheap money captured the imagination of both contemporaries and later writers, virtually all of whom have stressed, though with varying degrees of emphasis, the contribution of cheap money to the recovery of the 1930s. Pollard, for example, tends to play down the role of cheap money, the direct effects of which he feels could not have been large even in the building sector.[5] Sedgwick, on the other hand, while confessing ignorance about the transmission mechanism associated with shifts in nominal and real rates, finds it 'difficult to resist the conclusion that low rates were a very important element in the strong recovery of the 1930s.'[6] It would be strange had cheap money not had some beneficial effect on the economy but it is no doubt easy to exaggerate its role. After all, there was a convergence of several favourable factors at this time — devaluation, tariff protection, capital controls, an improvement in the terms of trade, rising real wages, shifts in consumer tastes and improved business confidence through an upturn in profits, that it is obviously difficult to isolate the causal influence of one single factor. However, the analysis which follows on the main channels of influence — bank lending, industrial investment and the new issue market, and the building industry — would suggest that some caution is required when assessing the contribution of cheap money to recovery.

As the economy recovered one would expect to see an expansion in the accommodation granted by the banks to industry and commerce, primarily for short-term purposes. With rising deposits for most of the period (an increase of over 16 per cent between 1933-37), the banks were in a relatively strong position to grant advances, while cheap money made possible a fall in the cost of borrowing and easier credit terms. Yet though the banks pursued a liberal credit policy during the 1930s, total bank advances continued to fall even after 1933 when the Bank of England's open-market operations had considerably enlarged the cash base of the banking system, and it was not until 1936 that there was a marked upturn in advances. By the end of the decade total advances had barely attained the level reached in the previous cyclical peak of 1929. Consequently, the ratio of advances to deposits fell significantly, from 55.5 per cent in 1929 to 39 per cent in 1936, subsequently rising to 44.1 per cent in 1939.

An analysis of advances by sector shows some quite sharp contractions between 1930-36. The decline in advances outstanding amounted to 50 per cent in the case of textiles, leather, rubber, chemicals, and miscellaneous and retail trades, by some 40 per cent in the case of heavy industries and mining and by one quarter for financial institutions, including building societies. The only sectors showing significant increases were the building trades and 'other advances' comprising professional and private loans and overdrafts. Even as late as 1938 most branches of activity, apart from construction, the entertainment trades and 'other advances', recorded a contraction in outstanding bank accommodation compared with 1929.[7]

There are several reasons why this unexpected trend, especially in the industrial sector, should have occurred. Some of the older declining industries were liquidating previous advances and because of their limited potential for expansion they required less in the way of financial accommodation. The fact that the banks became more selective in their lending policies, discriminating in particular against the more dubious debtors following the salutory experience of their over-generosity in the previous decade, would clearly tend to reinforce this trend. Secondly, some borrowers were finding it possible to secure accommodation outside the banking system, that is from insurance companies and other financial institutions. Thirdly, many firms were no doubt in a better position, when recovery got under way, to finance expansion from their own internal resources, a point which seems confirmed by the trend in new capital issues (see below). In this respect an important consideration was the alternative cost of finance to bank accommodation. Though interest rates generally fell sharply in the early 1930s, bank charges declined only slowly and the traditional floor to loan and overdraft rates prevented them falling much below 4½ per cent, though by 1935 rate differentiation had become quite common with the larger and safer borrowers obtaining rates as low as 3½ per cent. But for smaller and less fortunate firms the cost of borrowing from the banks proved to be only marginally lower than it had been in the 1920s except for the short periods when Bank rate was above 5 per cent.

It has frequently been argued, however, that within a certain range borrowers are relatively impervious to the cost of finance and that what really matters is whether it is readily available. This condition only holds of course when the rates and terms of finance are uniform throughout the broad spectrum of alternative sources. But in the 1930s this latter condition was not fulfilled. Given the stickiness of bank loan and overdraft rates it was generally cheaper for firms to reduce their bank debt and seek accommodation elsewhere, either by using undistributed profits, replacing bank debt by issued capital or by selling their holdings of securities, mainly in the form of government stock. This last solution was especially important. With the fall in interest rates gilt-edged prices were high and their yields low (about 3 per cent), and it was obviously sensible for firms to sell their securities at a capital gain to provide their financial needs rather than to use these assets as collateral against bank charges carrying a charge of around 4½ per cent. It is not surprising to find therefore that industry sold a considerable volume of gilt-edged stock in the 1930s, a large part of which was absorbed by the banking system, the investments of which more than doubled betwen 1931-36. Thus indirectly the banks became a source of finance for industry. That the banks should have been absorbing gilts when yields were low, when one would have expected them to have been releasing their own holdings on a buoyant market to finance their more profitable lending activities, is something which can only be explained by the unusual financial conditions of the time. The banks were forced into this position as a result of the ease and cheapness with which finance could be obtained from other sources. Banks faced competition from other financial institutions such as insurance companies, while industry found it cheaper to reduce its bank debt by selling securities or by placing new issues on the market. Thus the banks found their industrial advances dwindling during the 1930s while their investments rose. Whereas in more normal times they were prepared to hold some 55-60 per cent of deposits in the form of advances to industry, they could seldom attain more than 40 per cent in the 1930s, and were forced to hold the remainder in the form of government securities, holdings of which rose from £262 million to £610 million between February 1932 and October 1936.

A possible corollary of this situation is that the capital market played a more significant part in financing recovery. The more favourable monetary and financial conditions would tend, other things being equal, to encourage industrial investment and facilitate capital financing through the market. Cheap money would tend to push up share prices on expectations of improved profits which would facilitate equity financing by increasing the demand for securities. Secondly, the fall in interest rates would lower the cost of borrowing on debenture loans and preference shares, and also the equity costs of financing insofar as stock prices rose. Thirdly, the rise in gilt-edged prices and fall in yields would induce investors to move out of these securities and into industrial equities, which intially offered higher rates of return and the prospects of a capital gain. This in fact is more or less what happened after 1932. Share prices, as previously noted, moved up strongly from the trough of

1932 while the price of consols rose by no less than 29 per cent. Conversely, the yields on various classes of securities fell sharply between 1931-36 as the data in Table 1 indicate. The effect of these changes was as follows:

*Table 1: Security Yields and the Cost of Bank Accommodation 1931-36*

| | | Industrial Securities | | | |
|---|---|---|---|---|---|
| | 2½% Consols | Debentures | Preference | Ordinary | Cost of Bank Accommodation |
| 1931 | 4.4 | 6.2 | 6.8 | 8.03 | 5-6 |
| 1936 | 2.9 | 4.0 | 4.0 | 4.50 | 4-5 |
| Percentage fall | 34.1 | 35.5 | 41.2 | 44.0 | 10-25 |

*Note:* The figures for industrial securities include returns on both old and new issues.

*Sources:* E. Nevin, *The Mechanism of Cheap Money: A Study of British Monetary Policy, 1931-1939* (1955), pp. 206, 215 and London and Cambridge Economic Service, *Key Statistics of the British Economy, 1900-1966*, p. 16.

The low gilt-edged yields made industrial securities more attractive and so investors tended to switch out of gilts and into the latter. Insurance companies and investment trusts, for example, increased their holdings of industrial equities at the expense of government stock. Secondly, the release of investment funds and the decline in the cost of borrowing, especially compared with bank overdrafts, made it easier and cheaper for firms to acquire finance through the capital market.

Taking these considerations into account one would expect to see a rising tide of new issues through the market. Yet in fact the total of new capital issues (excluding government securities) never regained the 1929 level during this period. Moreover, when interest rates were at their lowest, 1933-36, the average amount raised annually was £171 million as against £236 million in 1930 and £311 million between 1927-29, or 80 per cent more than in the years 1933-36. However, for purposes of assessing the importance of industrial financing through the market, the aggregate figures are not very illustrative. They include capital raised by public and semi-public undertakings whose investment activity was not particularly sensitive to shifts in the cost of borrowing. The railways and electricity utilities are prime examples in this respect. Secondly, the aggregate data also include foreign borrowing by governments and institutions which collapsed to insignificant proportions in the early 1930s due to depression and increasing uncertainty in international markets, exchange controls, currency depreciation and the embargo on foreign lending. Thirdly, there is inevitably a certain amount of double counting in the figures insofar as they include issues by financial and investment trusts from which some of the money raised was used to buy industrial securities. Finally, some account needs to be taken of the changes in the price level during the period.

Rather more useful for our purposes here is the Midland Bank series for production, trade and industry which excludes issues by public boards, overseas borrowers and financial institutions. When deflated it provides a better indicator of real activity by industry on the stock exchange. It shows that new industrial issues fell from 100 in the later 1920s to 32 at the trough in 1931-32, and then rose steadily to 111 in the mid 1930s, after which it declined sharply again to 61 in the recession of 1937-38.[8] This trend scarcely suggests that industry was bursting to take advantage of the more favourable climate to raise money through the market. Only by the middle of the decade were new industrial issues higher in real terms than in the peak of the previous decade, and the sharp setback in 1937-38 cannot be explained by a significant worsening of money market conditions. Since investment in manufacturing was running at a higher level than in the previous decade — £86 million annually between 1933-38 compared with £68 million between 1924-29 (at constant 1930 prices) — it suggests that a larger proportion was being financed from undistributed profits. Balogh, for example, claimed that undistributed profits were more important than new issues in financing enterprise and that the expansion of some of the larger firms in the newer industries was based on own funding.[9] But even if correct this does not rule out the influence of cheap money altogether. Had interest rates been higher the more marginal investment projects would have been stillborn either because of the difficulty of raising new money to finance them, or because industrialists found a more profitable outlet for their cash reserves in the securities market. The very fact that industrial firms were disposing of their gilts to provide funding would appear to indicate that cheap money was having a favourable impact. Moreover, low interest rates also enabled companies to redeem some of their existing debt contracted at rates above those then ruling, which they did on a considerable scale. Between 1933-36 industrial borrowers converted £183.5 million into securities bearing a lower rate of interest, yielding an annual saving of over £2 million.[10]

Much of the evidence seems, therefore, to point to a rather muted response of industrial investment to the easier financial climate. In contrast to residential construction, domestic investment in manufacturing reacted rather belatedly to the upturn; not until 1934 was there a strong upsurge, which was then sustained through to 1938. Both investment and output in manufacturing rose to new peaks in the later 1930s, but whereas output growth was common to most sectors, the investment boom was confined to the metal and metal-using industries, where the pressure of demand provides a more powerful determinant of investment movements than lower interest rates. In the rest of manufacturing investment was only modestly higher by 1937 than in the previous cyclical peak of 1929. The heavy industries in particular responded late because of excess capacity constraints and when they did it was under the influence of rearmament rather than easier monetary conditions.[11]

There seems to be a good case for regarding the rising industrial investment as an acceleration response to increasing output and improved profitability rather than a product of cheaper money.[12] This conclusion squares with the findings of investment studies for the period which indicate a lack of sensitivity of non-housing investment to interest-rate changes,[13] thus confirming the conclusions of the less sophisticated Oxford inquiry of the late 1930s.[14]

The above conclusions do not of course apply to building or residential construction which is something of a special case requiring separate treatment. Building is not only a very important sector of the economy, accounting for a large part of total fixed investment, but it has also been accorded a special role in the recovery process of the 1930s (see Chapter 6). In addition, it is frequently cited as one of the chief beneficiaries of cheap money, being a sector relatively sensitive to changes in interest rates. Here again, however, there has probably been a tendency to exaggerate the role of cheap money in the building boom of the 1930s.

One thing should be made clear from the start: that cheap money affected only one segment of building work, namely houses built by private enterprise. It had little effect on local authority house building since this was determined largely by subsidy and slum clearance policies and in any case accounted for a very small proportion of the houses built in the 1930s. Nor is it particularly relevant to investment in industrial and commercial property (or for that matter public buildings), which did not rise significantly until the mid-1930s, and was motivated more by growth and profit prospects rather than by low interest rates. Nevertheless, this still leaves a very important segment, since private-enterprise housing constituted a large proportion of building activity in this period. Of the 2.5 million unsubsidised houses built privately in Britain in the interwar years, 1.8 million or 72 per cent of the total were completed between 1932-39. Over the same period investment in private-sector dwellings accounted for 54 per cent of all building investment and 27.5 per cent of total gross domestic investment. Building and related trades were also responsible for some 30 per cent of the increase in total employment in the first half of the 1930s.

As far as residential construction is concerned, the slide in interest rates had two main effects. First, it became cheaper to build and purchase houses, and second, investment in property became a more attractive proposition. Building firms, who relied heavily on short-term credit, found it easier and cheaper to get financial accommodation and, as noted earlier, this was one of the few sectors which made increasing use of bank accommodation in the 1930s. From the purchaser's point of view it became easier to buy a house because of the fall in carrying charges on new mortgages, from 6 to 4½ per cent, and a liberalisation of the building societies' lending terms. For the average house purchaser this could mean a reduction of up to one quarter or more in the cost of weekly payments. Fortunately, the building societies were able to satisfy the demand for new mortgages in the 1930s since they were literally awash with funds, the reason for this bonanza being that the shift in relative interest rates made them a safe and attractive form of investment. The yield on building society shares, which in the 1920s had been identical to that on long-term gilts, fell less steeply than the latter after 1931, so that during the period 1932-37, there was an average yield differential of 0.5 per cent in favour of building society shares. Finally, investment in residential property for letting became more attractive in this period as rents rose and gilt-edged yields declined.

Most authorities seem to be in agreement that cheap money exerted its strongest impact on the housing market, though there are differences of emphasis. There is no question that it made house-buying cheaper and easier and for the first time many working-class families were buying their own homes (before 1914 most accommodation had been rented). However, there are two important points to bear in mind when assessing the contribution of cheap money to the housing boom. In the first place, the downward shift in interest rates does not correspond very closely to the initial upswing in activity. In fact the upturn in unsubsidised house building began in the late 1920s and between the year ending March 1929 and March 1931 the number of such houses built rose from 66,000 to 131,000. This was followed by a temporary plateau in 1931-32 as a consequence of the depression and financial crisis, but in the following year (1932-33) output rose to 149,000. Clearly, therefore, there had been a very sharp rise in house construction before the easier credit conditions began to bite. It is true that there was a further big surge in private housing building in the next two years, 1934-35, when some 292,000 were constructed, and this steep rise may have been stimulated in part by improved borrowing terms and falling mortgage rates, though it should be noted that these were not at their best until the mid to late 1930s. However, private house building peaked at this point and in the later 1930s it actually declined.

The second point to bear in mind is that lower mortgage rates were only one of a number of factors which favoured residential construction in this period. On the supply side there was a severe shortage of housing accommodation at this time, accentuated by the lag in house building during the war, the rent controls of the 1920s which reduced the supply of property for letting, and the obsolescence of many Victorian properties, while the rise in the rate of family formation in the previous decade as well as the increasing migration of population between regions and to the suburbs accentuated the shortfall. Estimates of the shortage vary substantially, from 600,000 to 2 million by 1930.[15] Whatever the true figure, there was clearly a strong pent-up demand for housing ready to be unleashed by the early 1930s. Secondly, lower interest rates were associated with, or preceded by, a decline in building costs, rising real incomes, a reduction in Treasury subsidies for house building and a shift in consumer preferences, all of which helped to boost the private sector of the market. Probably the first of these factors is most relevant to the initial upswing in house construction. Marian Bowley's estimates indicate that the carrying charges on house purchase started to decline in the late 1920s and that between 1930-32 there was a fall of nearly 12 per cent. This was caused primarily by the fall in the capital costs of construction due to the decline in the cost of materials and labour which had begun in the late 1920s. Overall the aggregate index of building costs fell by 10 per cent between 1930-33.[16] The significant fact is that this fall took place before the new structure of mortgage rates had been put into effect and that it was sufficient to reduce the weekly cost of house-buying by rather more than a one percentage drop point in the mortgage rate. Moreover, there was an additional advantage to be derived from the fall in the cost of houses, in that it reduced the initial deposit required on a building society loan, an important consideration for people with small means. Since these cost changes occurred before lending rates and conditions had been relaxed, it can be argued that lower building costs rather than cheaper money were primarily responsible for the inception of the housing boom, a boom which was undoubtedly strengthened after the middle of 1933 when lower interest rates and more liberal lending terms gradually became effective.

Elsewhere in the housing market the impact of cheap money was even weaker. It does not appear that lower interest costs stimulated builders to construct houses on 'spec' to any significant extent, though speculative building was more prevalent than in the 1920s. Since credit costs did not form a very important part of the total costs of house construction the changes in the cost of finance did not have a significant impact on the volume of work done. The improvement in conditions of the rented market due to a relaxation of rent controls did encourage more investment demand for housing, though less than one third of the houses built between 1933-39 were for letting.

On balance therefore, cheap money should not be seen as a crucial element in the recovery of the 1930s, at least not in the early stages. As Arndt pointed out, cheap money, while beneficial, could not by itself decisively counteract deflation or produce recovery. What mattered in the final analysis was profit expectations for 'so long as entrepreneurs expected to make losses rather than profits on any new investment, interest rates could never be reduced sufficiently to make new investment appear worthwhile.'[17] The evidence suggests that cheap money was least effective in terms of bank lending and most influential in the housing market, while in the case of industry's call on the capital market it was neutral and unspecified. These broad conclusions require some qualification however. As we have seen, the banks indirectly assisted industry by absorbing their gilt-edged securities, whereas conversely, cheap money was only one of a host of favourable influences on the housing market. As yet, however, the precise quantitative impact of cheap money still remains unspecified. Counterfactually, one can no doubt argue that the recovery would have been less vigorous and sustained had interest rates not come down since in real terms they would have been very high. On the other hand, from a commercial point of view what often matters most is the ease with which credit and finance can be secured rather than its costs, provided of course that expectations about profitability are favourable. The contrast with the previous decade is often referred to in this context, but it would be unwise to stress the differences too strongly since there is no clear evidence that industrial developments were held up through lack of finance in the 1920s. Probably the biggest difference between the two periods lay in the greater degree of stability and continuity in the 1930s, the fair degree of certainty that credit conditions would remain favourable, in contrast to the volatile situation in the 1920s, and that profit expectations would continue to improve. As Morton noted: 'throughout the period 1932-39 the expectations of investors were continually conditioned by the published beliefs of government and monetary authorities that cheap money would continue. Thus the belief in cheap money helped to make it effective'.[18] On balance, cheap money was probably the most influential of the policy factors in this period, but in view of the marginal nature of many of the latter, that is not a very high commendation.

## The Fiscal Record

The government's fiscal policy in the 1930s has received ignominious slating by latter-day Keynesians. This is largely because it was never deliberately used for employment generation purposes. In contrast to a number of other countries, notably Sweden and America, where budgetary deficits were incurred, the British authorities stuck to the orthodox financial practice of budgetary balancing until late on in the period when rearmament expenditure drove the budget into deficit. Thus far from actively promoting recovery and employment, the fiscal stance tended to act in a perverse manner.

Despite this negative outturn, an analysis of fiscal policy can be quite instructive. The fiscal stance did vary somewhat during the period, with important leverage being imparted later in the decade, and modern analysis sheds greater light on the fiscal impact than conventional budgetary accounting. Secondly, it is worthwhile examining some of the reasons for the failure to adopt a countercyclical fiscal policy. And thirdly, there is the question of whether a full employment policy would have been a viable proposition at the time, given the economic constraints then prevailing.

One thing is clear, that at no time during the depression did the government contemplate an active fiscal policy: in fact rather the reverse. Fearful that falling tax yields and increasing social expenditure, because of mounting unemployment, would unbalance the budget and undermine confidence in Britain's financial soundness, retrenchment became the order of the day. In 1930 Snowden raised income taxes and indirect taxes, including those on beer and petrol. But much worse was to follow in the notorious economy campaign of 1931. At the end of July of that year the May Committee published its gloomy report on expenditure, which contained an alarming estimate of the budgetary deficit likely to arise unless stringent economies were adopted immediately. They proposed raising an additional £24 million by way of taxes and cutting expenditure by £96 million, the main burden of which was to fall on unemployment relief, to meet an estimated budgetary deficit of £120 million. In the event the proposed expenditure cuts split the Cabinet and helped to bring down the second Labour Government, and it was under the new National Government (August 1931) that Snowden, as Chancellor, introduced many of the May Committee's suggestions. In his autumn budget, indirect and direct taxes were raised significantly as were unemployment contributions. To contain expenditure the salaries of public employees, including teachers and civil servants, were cut, reductions were made in appropriations to the sinking fund, while transfer payments were reduced, including unemployment benefits. At the same time a stringent economy campaign was launched on the local authorities in an effort to get them to curtail spending, including that on relief works. The work of the Unemployment Grants Committee, set up at the end of 1920 to assist local authorities with work schemes designed to relieve unemployment, was brought to an abrupt halt. Though modest in its contribution to the unemployment problem — relief works of all kinds probably never accounted for more than 150,000 at any one time — it seemed singularly inappropriate to curtail its activities at the bottom of the depression. Henceforth, government job creation work was even lower, at best 59,000 being employed on government projects with a low of 303 in 1937.[19]

In the event the budget was almost balanced though the full effects of the measures did not really occur until 1932-33. The result was that total government spending (central and local) in real terms continued to rise thoughout the depression, by some 13 per cent (rather more for central government only), thereby raising its share in national income from 24 per cent in 1929 to nearly 29 per cent in 1931 and 1932, after which it declined.[20] On this reckoning therefore, government spending was exerting a stabilising influence, albeit a very mild one, during the depression itself. Moreover, though the budget was more or less in balance at a time when domestic conditions dictated that deficit financing would be more appropriate, the psychological benefits of financial rectitude have been stressed on several occasions. The restoration of financial soundness helped to restore confidence at home and abroad.[21] Another possible benefit recently put forward by Beenstock *et al.* is that the tightening of fiscal policy reduced public-sector borrowing and this in turn helped to bring down real rates of interest. If so then 'there was negative "crowding out" and the contractionary fiscal policy helped to pave the way for an investment boom. Thus, instead of the counter-cyclical fiscal policy exacerbating the recession, it had the opposite effect suggested by some present day commentators.'[22]

Even if somewhat belated, it should have been possible to start on a more reflationary course by 1932-33 in order to boost the economy. By then economic conditions had stabilised and confidence in sterling and Britain's financial position had been restored. After 1932 the budgetary situation was much easier as a result of rising tax yields, the new revenue from the tariff and the reduction in unemployment relief. Other constraints were also in the process of being relaxed. The deflationary impact of debt reduction and funding was eased considerably by the War Loan Conversion to a lower rate of interest in 1932, saving some £31 million on debt servicing costs, with later conversions and the fall in interest rates eventually resulting in debt servicing economies of around £100 million. The abandonment of the gold standard, the introduction of tariff protection and the establishment of the Exchange Equalisation Account also meant far less constraint from the external side than had been the case through to 1931. Finally, policies abroad, particularly in Sweden and the United States, demonstrated that unorthodox budgetary policies did not automatically lead to financial disaster. In short, once confidence in sterling and Britain's financial solvency had been restored, there did not appear to be any obvious reason for not adopting a more reflationary course of action.

Yet over the next two years little attempt was made to relax the fiscal stance. In fact, between 1932-34 central government spending in real terms fell quite sharply, while local authority spending remained static. Not until the budget of 1934 did the Chancellor begin to ease the fiscal burden. This reduced the standard rate of income tax by 6d. to 4s. 6d. in the pound, the unemployment benefit levels were restored while half the cuts in government salaries were rescinded, the other half being made good in the following year. In

1935 government spending started to rise again and from 1936 rearmament expenditure began to exert an increasingly significant impact on the budget. Between 1935-38 total government spending in real terms rose by £200 million, much of it being accounted for by the expansion of the rearmament programme. By the latter date defence spending accounted for 30 per cent of the total compared with 11 per cent in 1934. In current terms defence expenditure rose from £119 million in 1934 to £255 million in 1937 and £473 million in the following year, by which time it accounted for 9 per cent of national income as against less than 3 per cent in 1934.[23]

The magnitude of defence spending in the later 1930s forced the government into running budgetary defecits. The timing of the increase was quite fortuitous given the downturn in 1937-38. Several writers have stressed its importance in counter-cyclical terms in the later 1930s. 'There can be little doubt', wrote Bretherton and his co-authors, 'that defence expenditure was one of the major factors which put the brake on the beginning of a cumulative downward process and helped to stabilize activity at a level only moderately below that of the previous peak.' They estimated that a balanced budget in 1938 would have resulted in unemployment of around 3 million compared with the recorded figure for the year of 1.8 million.[24] More recently, Mark Thomas has shown even greater enthusiasm for the benefits to be derived from defence spending. He argues that the rearmament programme generated about one million man years of employment as against an overall unemployment total of nearly 2½ million at the start of the rearmament exercise in 1935. The major beneficiaries were coal, engineering, and iron and steel, especially the latter where one third of the emplyemtn growth between 1935-37 can be attributed to increased defence activity. At the same time defence spending also favoured some of the more depressed regions. The experience, he feels, demonstrates clearly the potential for counter-cyclical operations at an early stage of the recovery process. 'The success of rearmament in creating employment, even at the top of the cycle, leads us to view the eschewment of fiscal policy in the thirties as a missed opportunity for the economy.'[25]

Recent attempts to measure the aggregate impact of the fiscal stance by calculating the constant employment budget balance for the period suggest that the fiscal position of the early 1930s was even more restrictive than conventional budgetary analysis previously indicated. While the actual ex-post balance was fairly neutral through 1929-38, with a slight relaxation during the depression, the trend of the constant employment balance shows a consistent tightening of the fiscal stance through to 1933-34, the surplus amounting to 3-4 per cent of GDP. Thereafter there was a steady relaxation of some one per cent of GDP a year and by 1937 the constant employment surplus had been eliminated, to be followed by increasing deficits as rearmament expenditure expanded rapidly.[26] It should be noted however that the figures given in Table 2 cover only central government and social insurance funds and exclude local authority accounts. Secondly, they make no allowance for the differential leverage effects of various items of expenditure, nor have they been adjusted for price changes. Attempts to rework the data, by taking account of diffential demand effects and correcting for price changes to give the real budget surplus, have produced somewhat conflicting results, ranging from a stabilising fiscal stance through to 1933 to a marked and increasing tightening after 1931 which was not reversed until 1938.[27] It would appear therefore that a comprehensive assessment of fiscal policy in the 1930s still eludes us.

If much remains to be done to clarify the exact position of the fiscal stance during the 1930s, one thing is clear: that at no time did the government contemplate, let alone implement, a reflationary fiscal package to promote recovery and reduce unemployment. Two questions are worth pursuing at this point: Why did the government fail to use fiscal policy constructively, and how feasible would a full employment policy have been?

*Table 2: Constant Employment Budget Surplus (per cent of GDP)*

| | | | |
|---|---|---|---|
| 1929-30 | +0.4 | 1935-36 | +2.0 |
| 1930-31 | +1.1 | 1936-37 | +0.8 |
| 1931-32 | +2.5 | 1937-38 | 0.0 |
| 1932-33 | +3.0 | 1938-39 | -1.6 |
| 1933-34 | +4.2 | 1939-40 | -12.2 |
| 1934-35 | +3.2 | | |

*Source:* R. Middleton, 'The Constant Employment Budget Balance and British Budgetary Policy, 1929-39', *Economic History Review*, 34 (1981).

## Reasons for Fiscal Failure

The first issue has been debated endlessly in recent years, in part with a view to lifting some of the responsibility from the Treasury which has often been seen as the main stumbling block to action on the fiscal front because of its adherence to outmoded intellectual concepts.[28] It would be easy at this stage to argue that one could hardly expect radical experiments in fiscal policy in a world still dominated by classical economic thinking which preached the doctrine of sound finance, when concepts such as the constant employment budget balance were unknown, and when the Keynesian revolution in economic thinking was yet to provide the intellectual basis for deficit financing. It is true that conventional economic wisdom still acted as a barrier to the acceptance of new ideas and policies, but at the same time it would be wrong to assert that the absence of Keynesian analysis in its final form was the main reason for the failure to adopt a more constructive fiscal policy or a more active public-works programme. After all there were contemporary precedents for experiments with unorthodox techniques. As Arndt wrote in the 1940s: 'The most striking fact about British internal economic policy during the decade, however, is that Great Britain was the only one of five countries with which we are concerned which did not resort to a policy of budget deficits to promote internal recovery.'[29] He was referring to the United States, Germany, Sweden, and France (presumably under the Blum government), all of which experimented with new techniques. Sweden in particular had adopted Keynesian ideas before they had been presented in their final form. As early as 1933 the government abandoned the principle of balanced budgets and announced its intention of using budgetary policy as an instrument of recovery.[30] The Finance Minister's speech of January 1933 is notable not only for its open declaration of an unorthodox budgetary policy, but also because it formally acknowledged the state's responsibility for promoting recovery: 'the budget is based on the assumption . . . that in Sweden there will be no spontaneous tendency towards recovery, except to the extent that the policy of the State will help to bring it about ... In seeking to achieve this object, the State's financial policy must obviously play an important part.'[31] Accordingly, a large programme of public works was implemented and the resulting budgetary deficit was financed by loans which were amortised when recovery was well under way. By the mid-1930s public works absorbed 15-20 per cent of the Swedish budget, after which expenditure on public works was sharply reduced.

The American Federal government's fiscal policy in this period is even more interesting, not so much for what it achieved but for the fact that it is possible to demonstrate, as Stein has done, that much of the fiscal revolution to 1940 — the acceptance of budget deficits, increasing public expenditure and the avoidance of tax increases to balance budgets — can be described without reference to Keynes.[32]

The second point to bear in mind is that many of the main strands in the new economic thinking were being worked out or anticipated, albeit in a crude form, in the debate on public works during the 1920s. Public works as a weapon to alleviate unemployment have a long lineage — at least back to Elizabethan times, if not Roman — though their implications from an employment generating point of view and for budgetary analysis were not fully appreciated before 1914. During the 1920s the Labour Party and certain segments of the Liberal Party gave more explicit attention to their potential. From these debates emerged the concept of the multiplier and the notion that public works should be deficit financed rather than funded through increased taxation. Much of the limelight in the debate was captured by Lloyd George and his Liberal followers who, from the early 1920s, advocated a large-scale programme of public works. The culmination of Liberal thinking came at the end of the decade with the appearance of the Liberal 'Yellow Book' in 1928, *Britain's Industrial Future,* which became the basis of Lloyd George's famous election manifesto, *We Can Conquer Unemployment*, in the following year. It called for a large-scale public works programme to relieve unemployment which, though later found to be inadequate to meet the needs of the time, secured the approval of Henderson and Keynes in *Can Lloyd George Do It*? The importance of these two publications lay in the fact that they recognised, at least implicitly, the concept of the multiplier and the need for deficit financing, though it was not until two years later, in 1931, that Kahn presented a formal analysis of the multiplier in the *Economic Journal.* In the same year the *Report of the Macmillan Committee on Finance and Industry* appeared, which contained an important Addendum signed by Keynes and five other members, outlining the way ahead. The authors claimed that monetary policy, on which the committee placed a great deal of emphasis, could not be expected to raise employment to a satisfactory level and they rejected the alternative course of forcing down wages. On the other hand, the direct and indirect multiplier effects of a programme of new capital investment would, it was argued, raise the level of employment and the authors dismissed as unconvincing the traditional objections to this course of action: 'We are impressed for many reasons with what seems to us to be the greater wisdom and prudence of concentrating public attention of constructive schemes for encouraging national development rather than on efforts to drive down the general levels of salaries and wages.'[33] Finally, we should also note that Oswald Mosley, Chancellor of the Duchy of Lancaster in the 1929 Labour Administration and one of four ministers led by Thomas charged with the unemployment programme, was advocating even more radical action to cure unemployment. He subsequently resigned over the issue and went on to form a new party.

But if steps had been made towards the concept of a new policy direction by the early 1930s, the main protagonists were still voices crying in the wilderness. And the voices belonged to personalities scarcely likely to endear themselves to the Establishment: the flamboyant and mistrusted Lloyd George who was rapidly fading into the policital background, the brilliant but erratic and vacillating Keynes and the politically wayward Oswald Mosley. However, by 1933 a far larger measure of support for the policies of the dissidents was emerging. Prior to Chamberlain's budget of that year a vigorous campaign was launched with the aim of prompting the government to raise aggregate demand. It began with a series of leaders in *The Times* calling for bold policies of reflation, and no less than 37 economists came out in support of expansionist finance and dual budgetary accounting. Even the popular dailies, *The Daily Mail* and *The Daily Express,* together with a number of MPs took up the cause, while Keynes himself once more graced the pages of *The Times* in an attempt to sway the powers that be.

Sadly it all fell on deaf ears. As we know, there was never any systematic attempt to develop a countercyclical policy nor to carry through a large-scale programme of public works, least of all in the early 1930s. As Richardson pointed out, 'deficit financing was scorned, and the most reflationary action — increasing expenditure out of loans — was not even considered.'[34] Snowden's quest for balanced budgets at all costs set the tone for the rest of the decade, at least that is until rearmament began to upset the calculations.

The inflexible attitude of the authorities to any form of fiscal expansion can in part be attributed to the entrenched intellectual position of ministers and officials which prevented them from contemplating, let alone implementing, radical departures from the orthodox line. Nevertheless, up to 1933 at least, there were strong economic reasons which could justify a policy of inaction. In the difficult conditions prevailing in the early 1930s a radical programme involving large-scale state spending was out of the question, since its associations with budget deficits, inflation and adverse balances would have been enough to shatter business and financial confidence completely.[35] One has only to recall how the prospect of a relatively small budget deficit, which was not premeditated, had given rise to a crisis of confidence in 1931 to appreciate the significance of this point. In that year, when Britain left the gold standard and financial markets the world over were in a state of turmoil and panic, there could be no thought of implementing any radical programme. The first priority, as Chancellor Snowden explained to the House of Commons in February 1931, was to ensure that the country's budgetary position was sound and that Britain maintained her financial reputation, otherwise the consequences for Britain and the rest of the world would be disastrous. Similarly, in the previous month, Sir Richard Hopkins of the Treasury, in a memorandum presented to the Royal Commission on Unemployment Insurance, had warned the members that 'continued State borrowing on the present vast scale without adequate provision for repayment would quickly call in question the stability of the British financial system.'[36] By the summer of that year the situation had become a good deal worse, especially following the May Committee's revelations about the state of the public finances. Thus less than a month before the departure from gold the Prime Minister, Ramsay MacDonald, in a letter to a member of the House, dwelt on the waning confidence in Britain's financial system and the reason why corrective action could not be delayed: 'It is clear that in the midst of the world depression, whatever its causes, fears have arisen abroad as to the stability of our credit and the Budget estimates have fallen short most seriously. If our financial stability is endangered and a run made on our financial resources, the consequences are too terrible to envisage. This makes temporary retrenchment inevitable and imposes some amount of common sacrifice.'[37]

Whatever contemporary intellectual predilections dictated therefore, the economic and financial conditions of the early 1930s were scarcely the most propitious moment for breaking with established orthodoxy. In the event there was no major disaster. Britain's financial institutions, unlike those of many other countries, emerged unscathed from the traumas of 1931, and the government's relentless pursuit of financial orthodoxy did much to restore confidence both at home and abroad. Thus insofar as the government's fiscal policy helped to restore confidence and paved the way for lower interest rates in 1932, it could be regarded as a positive contribution. This approach also met with approval in Treasury circles. In their view, lower interest rates were not only the main weapon for dealing with recession,[38] but they had the added attraction of easing the servicing burden of the national debt. As we have seen, the opportunity was taken to convert a substantial part of the debt to lower rates, thereby reducing the servicing costs appreciably. But debt service costs still remained relatively high by comparison with the prewar period (or even the 1970s for that matter) and the Treasury, once having secured a reduction in the interest burden, was in no mood to contemplate the possibility of having to repeat the exercise again at some future date should an 'irresponsible' government be inclined to launch a large programme of public spending. In any case, the prospects for funding any such venture would have been particularly difficult without a rise in interest rates, which would have exerted a deflationary impact on the economy. In practical terms, therefore, 'The size of the debt as a problem *per se* would weigh in the policy regardless of whether projected government borrowing were to be financed by Treasury Bills or issues of stock.'[39]

Once the immediate crisis conditions had passed and some relief from the constraints of the national debt and external account had been secured, the climate would appear to have been somewhat more favourable for bolder experiments in fiscal policy. But in fact the budgetary stance of Neville Chamberlain was little different from that of his predecessor, Snowden: he combined 'experience with orthodoxy' and regarded sound financial policy (balanced budgets) as a prerequisite for recovery and, above all, for the maintenance of confidence in Britain's economic and financial system. Chamberlain, moreover, emphasised the fact that sound financial practice had made it possible to lower interest rates, which he regarded as essential for economic recovery.[40] In the latter respect, Chamberlain had a point to his credit, but he probably exaggerated the contribution of cheap money, as well as the dangers of any departure from financial orthodoxy. By the mid-1930s a moderate fiscal boost through deficit financing would probably not have been found too harmful, though by then the urgency had been removed somewhat simply because a strong recovery was well under way. But even by then the conventional view continued to prevail: 'even after four years of retrenchment, economy was still seen as a most desirable virtue by all parties, with few exceptions, and deficit budgeting as the remedy of the theoreticians. Chamberlain himself was content to be carried along by the rising tide of economic prosperity, his contribution being the restoration of confidence by the balancing of his budgets; and certainly, in the two years after 1931, this was no small contribution.'[41]

It seems therefore that the canons of financial rectitude were so deeply engrained within society that any attempt to resort to unusual financial practices would have created renewed worries about Britain's financial integrity, the strength of the currency, fear of inflation and the prospects of a flight of capital, which would have brought back 'the nightmare of recession'.[42] No doubt some of these fears were exaggerated in the light of past experience, but in times of difficulty the precautionary instinct tends to prevail, thereby barring the way to enterprising action. Consequently, the Cabinet was content to rely on 'natural forces and the restoration of investor confidence not because anyone thought that was the quickest way to reduce unemployment, but because ministers believed that it was the approach which provided the best balance of recovery and stability, with the former having no priority over the latter.'[43].

One is still left with the impression, however, that ministers and civil servants retained an innate suspicion of government spending, state intervention or anything which departed from the liberal practices of the nineteenth century. To many, high spending was as bad as budget deficits. 'It is no part of my job as Chancellor of the Exchequer', stated Snowden in 1924, 'to put before the House of Commons proposals for the expenditure of public money. The function of the Chancellor of the Exchequer, as I understand it, is to resist all demands for expenditure made by his colleagues and, when he can no longer resist, to limit the concession to the barest point of acceptance.'[44] Snowden was perhaps the most extreme of the interwar Chancellors, having an almost divine-like faith in the virtues of financial stringency. It was perhaps unfortunate that he presided as Chancellor throughout the depression since no one, least of all MacDonald, could stand up to him. 'As long as Snowden remained at the Exchequer there could be no Mosleyite or Keynesian unemployment policy. And no one in the Government was prepared to stand up to Snowden, because in the end they were all trapped in Snowden's assumptions, however hard they tried to kick against them.'[45] This brand of orthodoxy was accepted almost without question by his successor Neville Chamberlain. He, along with most members of the National Government, felt that it would be the height of folly to upset the 'natural' forces of recovery by expansionary and interventionist policies which it was thought would threaten to weaken the restorative influence of the slump in purging the system of 'unsound' investment and unproductive practices.[46] This view was based on the belief that depression and unemployment were part of the cyclical adjustment process and that this could best be assisted by retrenchment in expenditure, rationalisation and the use of monetary policy. Behind this lay the assumption that crises were the product of over-investment in relation to the volume of savings, which forced up costs and prices and adversely affected profit expectations. Thus the subsequent recession provided an automatic method of adjustment. Since the downturn in activity was assumed to be caused by too little savings and high costs, it could not be combatted by raising consumption or investment (aggregate demand). The only solution was to reduce expenditure (which meant public expenditure) and investment, which would lower costs and lead to a revival in business confidence. Hence the rather narrow Treasury view about investment and the general disinclination to accept the multiple effects of increased investment spending. Worse still, the concentration of attention on the business cycle adjustment process led to a failure to recognise that unemployment was more than simply a short-term cyclical phenomenon.

As far as public works were concerned, whether deficit financed or otherwise, their utility was questioned on the grounds that state spending on such projects would not create additional employment, since any increase in public investment would lead to a commensurate contraction in activity by the private sector. The Treasury is known to have attached great importance to this doctrine, though by 1931 Treasury officials appeared anxious to convince the Macmillan Committee that they no longer held this view.[47] Despite this disclaimer, it seems fairly certain that this sceptical attitude towards public works or any other form of public relief expenditure dominated official thinking during the 1920s and for much of the 1930s. The government acknowledged this as the reason why grants for relief works were restricted in 1925 and undoubtedly it had much to do with the fact that elaborate public relief schemes failed to materialise at any time during the 1930s. Civil servants advised against relief works and politicians were only too ready to accept their advice. Sir John Anderson, permanent under-secretary at the Home Office and head of a special secretariat to deal with unemployment, told Ramsay MacDonald in a memorandum dated July 1930 that 'We are now reaching the limits of works which will conform to any reasonable standard of economic utility or development ... The abandonment of any criteria of economic development with the consequential expenditure of public money at this juncture would be disastrous in the shock that it would give to the confidence which it is essential to maintain if the country is to get the benefit of world recovery when it comes. [The Government] must sweep away ruthlessly any lingering illusions that a substantial reduction of unemployment figures is to be sought in the artificial provision of employment.'[48] This advice was clear enough and was accepted readily by the Cabinet; within less than a year the government's unemployment programme had been axed. A year later (1932) the Cabinet of the National Government expressed even greater doubts about the utility of public works: 'Most of those who have been in office during the last five years were agreed that, whatever the past attractions of a public works policy, its application had been in many cases ill-considered and its disadvantages now far outweighed such advantages as it might have once possessed ... Evidence has taught us that they [relief works] do less good in the direct provision of work than harm in the direct increase in unemployment by depleting the resources of the country which are needed for industrial restoration.'[49]

The Treasury attitude to public works and expansionist finance may have softened a little during the course of the 1930s, but at best it was only a marginal shift in outlook and not one sufficient to accommodate an expenditure programme of the size necessary to make a serious dent into the unemployment problem.[50] For the most part the Treasury tried to ensure that government spending was kept as low as possible so that it did not divert loanable funds away from private enterprise. Budget balancing was regarded as essential for financial stability which meant that the budgetary procedure was very much a cash accounting exercise with little relevance to macroeconomic management. Thus until rearmament became a major issue later in the decade and forced the Treasury reluctantly to modify its stance,[51] the doctrines of minimum government involvement and innate hostility to public works and unorthodox financial policies were never seriously challenged.[52]

Civil servants were not lacking in ideas either when it came to finding other objections to public works proposals. Administrative and structural problems of execution were frequently raised in defence of inaction. The practical difficulties involved in initiating and organising public works programmes had in fact been seen as one of the chief obstacles to their implementation by the signatories of the Addendum to the Macmillan Committee. Investment projects would have to be planned for some years ahead and would have to be coordinated with the investment plans of local authorities which was no easy task given the large number of local councils. More recently, Middleton has outlined in some detail the administrative problems involved in launching a large-scale spending programme, not the least of which centred upon the relationship between central government and the local authorities with respect to the planning and execution of public works.[53] This time however it was non-Treasury officials who were most vociferous in their objections. The Ministry of Transport, for example, was opposed to big public works projects on technical grounds. According to the Ministry's consulting engineer, Sir Henry Maybury, no more than £20 million of useful projects could be readily assembled, and spending in the first round (year one) was normally limited to 20 per cent (in fact in the 1924-25 programme it had taken two years to reach even this low level). The combined opposition of the Treasury and the Ministry of Transport proved impenetrable and in September 1930 the Cabinet decided that little more could be accomplished through public works. In this instance, according to Marquand, it was the resistance of the road engineers rather than that of the Treasury mandarins that won the day.[54] Somewhat later, in 1935, the Ministry of Labour could be found opposing any extension of public works on structural grounds because it believed that it was not possible to provide beneficial employment of the type required, that is by trade skill or area. The Ministry doubted whether more than one quarter of a million men of the type normally required for public works would actually be forthcoming given the wide variation in age, sex and physical capability of those registered as unemployed, together with the natural inclination among adult males to wait in the hope of being reabsorbed into their former occupations rather than venture into new work with limited long-term prospects. Furthermore, the Ministry of Labour maintained, and with some justification, that indiscriminate public works which failed to improve the structural problems of the basic industries and depressed regions would be positively harmful.[55]

Lastly one might note that neither the Labour Government nor the National Government was ever under severe pressure from the Labour movement in general to depart from orthodox policy. The trade-union movement had few constructive proposals for curing unemployment — work-sharing schemes were rejected and the craft unions tightened their restrictions on the entry of apprentices. It is true that the unions did support the idea of expanding public works but they failed to make much impression on the establishment partly because they were unable to make out a constructive and coherent case about the employment-generating effects of increased spending or the need for deficit financing. Moreover, after the fall of the Labour Government in the summer of 1931 the unions shifted their stance towards the wider issues of nationalisation, planning and rationalisation, none of which was particularly in tune with short-term reality.[56] In fact, once the unemployment trough had been reached and economic conditions began to improve, the tone of the unemployment debate became decidedly muted and by the middle of the 1930s the subject had all but disappeared from the trade-union agenda, including that of the TUC.[57] Instead, once the spectre of mass unemployment had receded somewhat, the trade-union movement began to pay more attention to those with jobs rather than those without. This change in outlook was reflected in the campaign for a 40-hour week, claims for higher wages and the resistance to any erosion of the value of unemployment benefits. The latter was regarded as a particularly important issue since it not only relieved the unions of the responsibility for supporting those out of work (as had been the case prior to the war), but it also helped to reduce the competition for jobs, which would have depressed wages. In other words the unions, perhaps with indecent haste, readily accommodated themselves to the new level of unemployment as the price to be paid for maintaining the wage levels of those in work, the corollary of which was to focus attention on improving rates of unemployment benefit rather than on expanding employment.[58]

Thus the government had little to fear from the Labour movement whose philosophy of maintenance rather than work creation accorded so well with established practice. That it might have done more to exacerbate the unemployment problem by maintaining the level of real wages (see Chapter 6) was beside the point to a government bent on preserving social harmony. Similarly, the National Unemployed Workers' Movement, set up in 1921, offered even less of a threat to the government. Despite its rallies, hunger marches and persistent agitation, it had few constructive proposals to make other than those aimed at redefining unemployment out of existence, by keeping youngsters at school longer and by earlier retirement. It too sought to protect the standard of living of those in work by refusing employment at less than union rates while at the same time campaigning for improved benefits.[59] Finally, Garraty mentions the social stigma attached to forcing the unemployed into public service by cutting off the dole as an obstacle to public works: 'The well-established unemployment-insurance system, buttressed by the dole (officially known by the euphemism "transitional payments") created a formidable political obstacle for those in favour of work relief in Britain: to compel the unemployed to accept public-service jobs by threatening to cut off the dole resembled the detested work-or-starve practice under the old poor law.'[60]

In short, the reasons for the failure to adopt a more active fiscal policy cannot be attributed to any one factor. A combination of economic circumstance, administrative intransigence and intellectual torpor (a preconception of what was right) prevented the move towards a more dynamic fiscal policy, at least that is until the pressure of international events forced the reluctant acceptance of the inevitable.

## What Price Full Employment?

Finally, in view of its relevance to the contemporary unemployment problem of the 1980s, we might justifiable ask whether a full employment recovery programme would have been feasible in the 1930s. Oddly enough, though the unemployment problem of the period has frequently been seen as one arising from aggregate demand deficiency requiring government intervention to restore equilibrium, there has been far less analysis of what this would have involved in practice, nor has there been an adequate appreciation of the constraints to a full employment policy. Thus in a recent disequilibrium analysis of interwar employment, Broadberry sees the structural problem of the period in terms of general demand deficiency arising from a too rapid contraction of certain sectors, so that any relative price changes signalling a transfer of resources from old to new sectors get swamped by the quantity signals emanating from deficient aggregate demand. The result is that the economy 'became locked in a position of chronic under-employment, with the depressed state of aggregate demand acting as a disincentive to invest in any industry, new or old.' While this statement begs several questions, the point at issue here is that we are not told how this disequilibrium position could have been unlocked, much less whether it would have been feasible to do so in the light of the constraining factors likely to arise from so doing. We are merely informed that government intervention is required.[61]

Neither Lloyd George nor Keynes, of course, anticipated that they could cure unemployment completely and their programmes fell far short of what would eventually have been required. Lloyd George's original programme of about £260 million worth of public works expenditure over a two-year period was estimated to lead to an additional employment of 611,000, while the Keynes/Henderson variant of an annual expenditure of £100 million for three years would produce employment for some half a million workers. Both however rather overestimated the multiplier effects of increased spending. Thomas has reworked the Keynes/Henderson variant of the Lloyd George programme over a five-year period using a long-run multiplier of 1.44 and finds a total employment impact of 359,000. This suggests that it would have had to be multiplied many times over for there to have been a worthwhile impact on the level of unemployment at the depth of the depression.[62] Using Thomas's estimates for the employment generation effects of new government spending one comes up with an upper bound requirement of £752 million of spending to deal with an unemployment of three million, while Glynn and Howells suggest a figure of £537 million.[63] Earlier estimates by Kaldor of the spending required to meet a target of 1.25 million unemployed for 1938 also indicated that a substantial public outlay would have been required.[64]

The sums required would clearly have been substantial, though they do not fully take account of the offsets to expenditure through savings on unemployment benefits and additional tax revenues as employment rises. Thus the final net cost to the Exchequer would have been reduced, though in the initial years, before the savings fully accrued, the total government outlay would have been of the order specified. In terms of the national income and public outlay of the period the sums involved are enormous. For example, the high projection of £752 million is almost equivalent to the total budgetary outlay in 1932 and no less than 19.3 per cent of GNP. Even Glynn and Howells' estimates are formidable, nearly 14 per cent of GNP, 49 per cent of total public authority spending, and involving an increase in government spending or reduction in revenue equivalent to 70 per cent of the national budget for 1932. Thus, even after allowing for the savings in later years, the initial impact in relative terms would have been horrendous, so much so that the authors remark that 'Even before one asks where the funds to meet the deficit might have come from, the required amount can already be seen to be in the realms of political and economic fantasy.'[65]

A programme of spending of the magnitude outlined was obviously a political non-starter in the 1930s. Its economic and financial implications would have given the Treasury nightmares, apart from the fact that it would have been impossible to finance it in a bond market where interest rates were already at rock bottom. As it was the Treasury found it difficult to market stock in the 1930s and any further funding would have pushed up interest rates sharply.[66] The domestic financial constraint is clearly a powerful one, as experience has taught us in more recent times. The second main difficulty would have been the strain imposed on the balance of payments. Thomas has calculated that even Lloyd George's modest programme would have produced an adverse shift in the current-account balance of between £20-25 million. A more ambitious programme could have led to a deterioration of £150 million or more, while Kaldor suggested a possible adverse balance of £130 million in 1938 as against an actual deficit of £55 million. Moreover, the combination of an adverse current-account balance and unbalanced budgets would have caused a capital outflow and weakened the capital account of the balance of payments.[67]

The strains imposed by rearmament spending in the later 1930s suggest that it was having an adverse influence on the balance of payments and on financial confidence. It led to a diversion of industrial capacity away from exports and a rise in imports so that the current account deteriorated. At the same time fears about the consequences of financing increasing budgetary deficits as a result of defence spending caused a weakening of the exchange rate which alarmed both Keynes and the Treasury. By the middle of 1939 the effects of rearmament spending on Britain's economic and financial stability were causing grave concern in official circles and it was doubted whether Britain would be able to stand the strain for very long.[68] Whether the external effects of a large fiscal thrust for employment purposes could have been alleviated by devaluing sterling from its enhanced effective rate is debatable. There is some doubt whether in fact devaluation would have been effective given the fact that the sum of the import and export price elasticities was slightly below unity.[69] Furthermore, if it is assumed that Britain had taken unilateral fiscal action, then presumably prices and costs would have risen relative to those abroad thus worsening Britain's competitive position.

In other words, there are both political and economic constraints to be considered in contemplating a counterfactual employment generating programme for the 1930s. The economic and financial constraints may not have been as serious as in more recent times, given the absence of serious inflation, but they existed all the same. Had a reflationary boost of any magnitude been adopted some restrictions would have been required to prevent the economy from short-circuiting, as Arndt recognised in 1944. Mindful of the balance of payments constraint arising from heavy spending, he felt this alone would have required recourse to a controlled economy along the lines of the German model: 'An expansionist policy would ... in all probability ... have meant the transformation of the British economy into a largely State-controlled, if not planned, economic system.'[70] It goes without saying that the Conservative-dominated National Government of the 1930s would never have acquiesced in such an interference with private enterprise.

One final point is worth mentioning. A generalised fiscal reflation would not really have solved the long-term structural problems of the depressed regions and the staple industries. It has long been recognised that macro-policy is too blunt an instrument to deal with such problems. Keynes, in his famous series of articles to *The Times* early in 1937, was aware that the time was approaching when 'there is not much advantage in applying a further general stimulus to the centre ... We are more in need today of a rightly distributed demand than of a greater aggregate demand.'[71] At the same time, sustained recovery and later rearmament were putting pressure on the more prosperous southern regions and certain sectors of the economy, which manifested itself in severe shortages of skilled labour in several occupations. Such shortages could exist amid abundance because of the relative rigidity of the economic structure and the lack of homogeneity in the labour market. The labour market in particular presented a difficulty because of serious mismatch problems; skill shortages could not easily be overcome by employing redundant workers because of their lack of adaptability and resistance to mobility, while union work practices made it difficult to economise on the use of skilled labour.[72] Moreover, Carol Heim has shown that organisational constraints set limits to the rate at which the depressed areas could have responded to a generalised fiscal thrust. She maintains that the barriers to re-entry into the national economy were strong because of conditions of controlled competition in industries most suitable for the regeneration of the depressed regions, while large-scale firms outwith these regions had not yet reached the stage where they were ready to disperse their activities in branch plants to the peripheral areas. 'In interwar Britain the organisational prerequisites were not yet in place to have enabled a vigorously expansionary macroeconomic policy to generate, in the short run, large numbers of branch plants dispersible to the depressed areas.'[73] On the other hand, a more discriminating approach centred upon specific regions and industries may not have been the best method in the long run if it had meant featherbedding sectors which were in decline.[74] Indeed, such a policy would have conflicted with such attempts as there were to rationalise these industries by eliminating their excess capacity (see Chapter 5), and in this respect the Treasury's response to Lloyd George's public works proposals, that should they be successful they would tend to encourage industry to struggle along instead of rationalising itself, contained more than a grain of truth.[75]

What we are saying, therefore, is that after a severe depression and prolonged period of structural imbalance, as in interwar Britain, it is difficult to mop up underutilised resources quickly and without cost. A rapid solution runs up against financial and economic constraints, and it does not necessarily solve the underlying structural problem. Indeed, insofar as the decaying sectors are given a new lease of life by reflationary action, it may simply defer the much-needed adjustment process. This does not mean that more could not have been done in the 1930s to tackle the twin problems of unemployment and structural decay, but the experience of the 1980s leads one to be cautious about the claims made for fiscal policy as a solvent for all problems.[76]

## Notes

1. *Report of the Committee on Finance and Industry,* Cmd 3897 (1931), p., 136.
2. In fact it has been argued that the rise in the money supply after 1932 was 'a consequence of the policy to cap sterling rather than a textbook application of the monetary independence vested in a policy of floating exchange rates.' M. Beenstock, F. Capie and B. Griffiths, 'Economic Recovery in the United Kingdom in the 1930s', *Bank of England Panel Paper,* 23 (1984), p. 64.
3. 'The 1932 War Loan Conversion Con Trick', *Investors' Chronicle,* 23 December 1983, pp. 15-16
4. It should be noted that the real money base did expand in the earlier period because of the fall in prices, but it grew even faster in the 1930s. Nominal money growth in the latter period was however less than nominal GDP growth.
5. S. Pollard, *The Development of the British Economy, 1914-1980,* (1983), pp. 151-2.
6. P. N. Sedgwick, 'Economic Recovery in the 1930s', *Bank of England Panel Paper,* 23 (1984), p. 43.
7. W. A. Morton, *British Finance, 1930-1940* (1943), p. 261.
8. The detailed figures can be found in Nevin, *op.cit.,* p. 222.
9. T. Balogh, *Studies in Financial Organisation* (1974), p. 278; of Nevin, *op cit.,* p. 248 and H. W. Richardson, *Economic Recovery in Britain 1932-9* (1967), pp. 201-2.
10. Nevin, *op.cit.,* pp. 227-8.
11. R.C.O. Matthews, C. H. Feinstein and J. C. Odling-Smee, *British Economic Growth, 1856-1973* (1982), pp. 384-5. The authors doubt whether the fall in interest rates had much effect on manufacturing investment.
12. G. D. N. Worswick, 'The Recovery in Britain in the 1930s', *Bank of England Panel Paper,* 23 (1984), p. 20.
13. P. J. Lund and K. Holden, 'An Econometric Study of Private Sector Gross Fixed Capital Formation in the United Kingdom, 1923-1938', *Oxford Economic Papers,* 20 (1968); D. J. Thomas and G. Briscoe, 'Investment and Capacity Utilisation in the United Kingdom, 1923-1966', *Oxford Economic Papers,* 23 (1971); T. J. Thomas, *Aspects of UK Macroeconomic Policy during the Interwar Period: A Study in Econometric History,* University of Cambridge, Ph.D. (1975).

14. See the papers by Henderson, Andrews and Meade in *Oxford Economic Papers*, 1 (1938), 3 (1940).
15. Pollard, *op. cit.*, p. 153.
16. See M. Bowley, *Housing and the State, 1914-44* (1945), p. 278 and I. Bowen, 'Building Output and the Trade Cycle (UK 1924-38)', *Oxford Economic Papers*, 3 (1940).
17. H.W. Arndt, *The Economic Lessons of the Nineteen Thirties* (1944), p. 256.
18. W.A. Morton, *British Finance, 1930-1940* (1943), p. 248.
19. J.A. Garraty, *Unemployment in History* (1978), p. 211.
20. The data are given in A.T. Peacock and J. Wiseman, *The Growth of Public Expenditure in the United Kingdom* (2nd edn. 1967), pp.159, 164-5, 202.
21. See Richardson, *op. cit.*, pp.218-9.
22. Beenstock, Capie and Griffiths, *loc. cit.*, p.64.
23. See G.C. Peden, 'A Matter of Timing: The Economic Background to British Foreign Policy, 1937-1939', *History*, 69 (1984).
24. R.F. Bretherton, F.A. Burchardt and R.S.G. Rutherford, *Public Investment and the Trade Cycle in Britain* (1941), pp.84, 92.
25. M. Thomas, 'Rearmament and Economic Recovery in the Late 1930s', *Economic History Review*, 36 (1983), pp.570-2.
26. R. Middleton, 'The Constant Employment Budget Balance and British Budgetary Policy, 1929-39', *Economic History Review*, 34 (1981).
27. See S.N. Broadberry, 'Fiscal Policy in Britain during the 1930s', *Economic History Review*, 37 (1984), and the reply by Middleton in the same issue; also Sedgwick, *loc. cit.*, p.47.
28. G.C. Peden, 'The "Treasury View" on Public Works and Employment in the Interwar Period', *Economic History Review*, 37 (May 1984).
29. Arndt, *op.cit.*, p.125. In all fairness the Labour Government which took office at the start of the depression did make a modest attempt to plan a compensatory public-works policy through the UGC, but it was on a totally inadequate scale and was hampered by the government's weak position, by planning difficulties and it ran foul of the Treasury and vested interests. Eventually it was cut short by the economic crisis and economy campaign in 1931.
30. D. Winch, 'The Keynesian Revolution in Sweden', *Journal of Political Economy*, 74 (1966) and *Economics and Policy. — An Historical Study* (1969) p.209. Developments in other countries are discussed in D.H. Aldcroft, 'The Development of the Managed Economy before 1939', *Journal of Contemporary History*, 4 (1969).
31. B. Thomas, *Monetary Policy and Crises* (1936), p.208.
32. H. Stein, *The Fiscal Revolution in America* (1969), p.131.
33. *Report of the Committee on Finance and Industry*, Cmd 3897 (1931), pp.203, 208.
34. Richardson, *op. cit.*, p.230.
35. Despite the sagging price level the inflationary implications of 'unsound' policies were not easily dismissed. As Overy comments in the German context, 'The link in people's minds between unorthodox economic policies and inflationary crisis may well have been a false one, but it nevertheless acted as a major psychological constraint in any discussion of new ways to fight the crisis.' R.J. Overy, *The Nazi Economic Recovery, 1932-1938* (1982), p.25.
36. R. Bassett, *Nineteen Thirty-One Political Crisis* (1958), pp.44-6.
37. R. MacDonald to E. Edwards, MP, 25 August 1931.

38. Peden, *Economic History Review* (1984), *loc. cit.*, p. 180.
39. J. D. Tomlinson, 'Unemployment and Government Policy between the Wars: A Note', *Journal of Contemporary History*, 13 (1978), p. 72.
40. B. E. V. Sabine, *British Budgets in Peace and War, 1932-1945* (1970), pp. 15-16, 299.
41. *Ibid.*, p. 74.
42. *Ibid.*, p. 299.
43. F. C. Miller, 'The Unemployment Policy of the National Government, 1931-1936', *The Historical Journal*, 19 (1976), p. 476.
44. C. Cross, *Philip Snowden* (1966), p. 207.
45. R. Skidelsky, *Oswald Mosley* (1975), p. 210.
46. W. R. Garside, 'The Failure of the Radical Alternative: Public Works, Deficit Finance and British Interwar Unemployment', *Journal of European Economic History*, (1985).
47. *Report of the Macmillan Committee on Finance and Industry*, pp. 203-4.
48. R. Skidelsky, *Politicians and the Slump. The Labour Government of 1929-31* (1967), pp. 216-17.
49. Quoted in J. Stevenson and C. Cook, *The Slump: Society and Politics during the Depression* (1977), p. 63.
50. A. Booth, 'The Keynesian Revolution in Economic Policy-making', *Economic History Review*, 36 (1983), p. 123; S. Howson and D. Winch, *The Economic Advisory Council 1930-1939: A Study in Economic Advice during Depression and Recovery* (1977); G. C. Peden, 'Sir Richard Hopkins and the "Keynesian Revolution" in Employment Policy 1929-1945', *Economic History Review*, 36 (1983).
51. G. C. Peden, 'The Treasury as the Central Department of Government, 1919-1939', *Public Administration*, 61 (1983), p. 375. Peden makes the point that the Treasury's supremacy in finance lasted only so long as the Cabinet was prepared to adhere to financial orthodoxy. Once war needs led to a relaxation then Treasury control over spending waned, since it was no longer necessary for it to ensure that spending departments kept within the limits set by the Cabinet's financial priorities.
52. Garside, *loc. cit.*; R. Skidelsky, 'Keynes and the Treasury View: The Case for and Against an Active Unemployment Policy, 1920-1939', in W. J. Mommsen (ed.), *The Emergence of the Welfare State in Britain and Germany, 1850-1950* (1981), pp. 185-6.
53. R. Middleton, 'The Treasury in the 1930s: Political and Administrative Constraints to Acceptance of the "New" Economics', *Oxford Economic Papers*, 34 (1982) and 'The Treasury and Public Investment: A Perspective on Interwar Economic Management', *Public Administration*, 61 (1983).
54. D. Marquand, *Ramsey MacDonald* (1977), pp. 552-3.
55. Garside, *loc. cit.*
56. S. Pollard, 'Trade Union Reactions to the Economic Crisis', *Journal of Contemporary History*, 4 (1969), pp. 108, 111-13.
57. *Ibid.*, p. 115.
58. J. A. Garraty, 'Unemployment during the Great Depression', *Labour History*, 17 (1976), pp. 135-8. In fact the TUC General Council in their evidence to the Macmillan Committee had argued in favour of maintaining the living standards of those in work at the expense of the unemployed. 'We would prefer to wait for international action, in the meantime pressing forward the re-organisation of industry while maintaining those unemployed, and preserving

the present standard of living for those in employment, rather than have unemployment eliminated immediately at the cost of a degradation in the standard of living of the workers.' Quoted in M. Casson, *Economics of Unemployment* (1983), p.183.

59. Pollard (1969), *loc. cit.*, p.114; Garraty (1976), *loc. cit.*, p.141.
60. J.A. Garraty, *Unemployment in History* (1978), p.207.
61. S.N. Broadberry, 'Unemployment in Interwar Britain: A Disequilibrium Approach', *Oxford Economic Papers*, 35 (1983), pp.464, 469, 483. The analysis deals inadequately with the supply side and does not explore the implications of any induced demand changes in terms of the structural problem.
62. T. Thomas, 'Aggregate Demand in the United Kingdom, 1918-45', in R. Floud and D. McCloskey (eds.), *The Economic History of Britain since 1700*, Vol. 2, *1860 to the 1970s* (1981), p.337.
63. D.H. Aldcroft, *Full Employment: The Elusive Goal* (1984), p.41; S. Glynn and P.G.A. Howells, 'Unemployment in the 1930s: The Keynesian Solution Reconsidered', *Australian Economic History Review*, 20 (1980), p.41.
64. Published as an appendix to W. Beveridge's *Full Employment in a Free Society* (1944).
65. Glynn and Howells, *loc. cit.*, p.42.
66. J.D. Tomlinson, 'Unemployment and Government Policy between the Wars: A Note', *Journal of Contemporary History*, 13 (1978), p.74.
67. Thomas thesis (1975), pp.226-7; Aldcroft, *op. cit.*, pp.47-9.
68. Peden, *History* (1984), *loc. cit.*, pp.17-18, 25, 27. In the financial year 1939 Britain was devoting almost as much of her GNP to rearmament as Germany and in the following year had well surpassed her. R.A.C. Parker, 'The Pound Sterling, the American Treasury and British Preparations for War, 1938-1939', *English Historical Review*, 98 (1983).
69. Thomas thesis (1975), p.228.
70. Arndt, *op. cit.*, pp.134-5; see also R.J. Overy, *The Nazi Economic Recovery, 1932-1938* (1982), p.51, where he points out that Nazi spending necessitated extensive controls over the whole economy.
71. J.M. Keynes, 'How to Avoid a Slump I. The Problem of the Steady Level', *The Times*, 12 January 1937.
72. See R.A.C. Parker, 'British Rearmament 1936-9: Treasury, Trade Unions and Skilled Labour', *English Historical Review*, 96 (1981); R. Shay, *British Rearmament in the Thirties: Politics and Profits* (1977), pp.125, 134; G.C. Peden, 'Keynes, the Treasury and Unemployment in the Later Nineteen-Thirties', *Oxford Economic Papers*, 32 (1980), p.15.
73. C.H. Heim, 'Uneven Regional Development in Interwar Britain', *Journal of Economic History*, 43 (1983), p.275 and 'Industrial Organisation and Regional Development in Interwar Britain', *Journal of Economic History*, 43 (1983), pp.931, 934.
74. Cf. J.F. Wright, 'Britain's Inter-war Experience', *Oxford Economic Papers, Supplement*, 33 (1981), p.298.
75. Skidelsky in Mommsen (1981), p.182.
76. A further recent note on this issue by Garside and Hatton argues the case for an expansionary fiscal policy in the 1930s in the belief that the constraints in the way of so doing have been exaggerated. But at the same time they recognise that there were obstacles to be overcome which seem to negate their original proposition. Nor do they deal with all the alleged constraints. In any

case both Thomas and Peden, and Arndt before them, have shown that a full-blown Keynesian programme would have required extensive controls (as in Germany) to prevent the economy from short-circuiting. W. R. Garside and T. J. Hatton, 'Keynesian Policy and British Unemployment in the 1930s', *Economic History Review*, 38 (1985); S. Glynn and A. Booth, 'Building Counterfactual Pyramids', *Economic History Review,* 38 (1985); T. Thomas, 'Aggregate Demand in the United Kingdom, 1918-45', in R. Floud and D. McCloskey (eds.), *The Economic History of Britain since 1700,* Vol. 2, *1860 to the 1970s* (1981), p. 337; G. C. Peden, 'The Treasury View on Public Works and Employment in the Interwar Period', *Economic History Review*, 37 (1984); H. W. Arndt, *The Economic Lessons of the Nineteen-Thirties* (1944), pp. 134-5. On Germany see R. J. Overy, *The Nazi Economic Recovery, 1932-38* (1982), p. 51, where he points out that Nazi spending policies were dependent on extensive controls over the whole economy to make the system work.

# 5 Regional and Industrial Policies

Given the severity of Britain's structural problems, which became more conspicuous in the 1930s, it was almost inevitable that the government would take some action to assist the more depressed sectors of the economy. In fact, there were quite a lot of policy measures at the micro level, though mostly of a rather *ad hoc* type, and in total they did not contribute a great deal to recovery. Regional policy was very limited in scope and provided only a small number of jobs. Industrial policy, if it could be called that, varied a great deal from one industry to another, and often the employment needs clashed with plans to try and rationalise capacity. Agriculture received the most attention and is dealt with separately.

## Regional Policy

The severe regional imbalance between North and South which emerged after the war became more pronounced in the 1930s. The old staple industries of the North fared badly in the depression, partly because of the large decline in exports, and even in the recovery phase the prospects for employment growth were bleak. As we saw in Chapter 1, the chief problem of the northern regions lay in their excessive concentration on a narrow band of heavy industries, the market for whose products collapsed largely, though not entirely, because of the severe deterioration in their export trade. In almost every year between 45-50 per cent of total unemployment was to be found in five areas — South-east Scotland, the North-east coast, Lancashire, Merseyside and South Wales — though these areas accounted for less than one third of the insured population and just over one quarter of the total population of the country.[1] Rates of unemployment in these regions were often more than twice the national average and for most of the time they were losing employment. Thus even by 1936, after several years of recovery, insured unemployment in the North of Britain was 18 per cent as against 7.4 per cent in the South, a greater disparity than in

1929. Moreover, the depressed regions also had a high proportion of long-period unemployment. In contrast, the South and Midlands, with a greater incidence of expanding industries and service trades, were less affected by depression and recovered more quickly after 1932.

The consequences were that population, employment and industrial activity tended to gravitate to the southern half of the country as the North became locked into a vicious circle of depression and decay, making it difficult to regenerate the old industries or attract new ones. Most of the net increase in factories occurred in the South-east, whilst Wales, Scotland and parts of the North of England suffered a net loss of factories as closures exceeded new openings. As a result, between 1924 and 1935 the southern and midland regions increased their share of net industrial output from 44.6 to 57.3 per cent, whereas there was a corresponding fall in that for the North, from 55.4 to 42.7 per cent.[2] Not surprisingly therefore, all northern regions lost population through migration, while Wales actually suffered an absolute decline in population during the interwar period.[3]

Little was done to alleviate the regional problem in the 1920s until late on in the period when a scheme of assisted migration and industrial training was inaugurated. Originally designed for the benefit of coalminers, the scheme was subsequently extended to assist and encourage workers in depressed areas to look for work in the more prosperous regions. It was administered by the Ministry of Labour through the employment exchanges, and altogether some 280,000 individuals benefited from assisted migration between 1928-38[4]. By way of comparison it is interesting to note that between 1922-33 over 400,000 Britons were placed in Commonwealth countries under the Empire Settlement Act of 1922, though how far the small sums received influenced their decision to migrate abroad must remain a matter of speculation.[5]

The unemployment problem in the depressed regions was so large and specialised that any scheme of assisted migration could only serve as a temporary palliative. Moreover, it did not actually create any new jobs, while the people most likely to take advantage of it were young and single workers who were more mobile and better able to grasp new skills than older workers with family responsibilities. Consequently, in the 1930s, the government supplemented its labour mobility scheme by a policy of bringing work to the workers, in what came to be known as the special areas. The basis of the special areas legislation 1934-37 was laid out in three Acts, the first of which (1934) established Commissioners for Scotland and England and Wales whose task was to promote the economic and social development of four designated areas: Central Scotland, South Wales, West Cumberland and Tyneside including parts of Durham. At first their powers and source of finance were fairly limited but these were extended in 1936 and 1937 to include the power to attract new industries and firms to the areas by financial assistance, to contribute to rent, rates and taxes of firms setting up there, to provide factory premises, and to establish trading estates. Finally, foreign firms were to be encouraged to locate in these areas, while the government would favour firms in these areas when awarding defence contracts.

Though a start in the right direction, regional policy could scarcely be described as a resounding success. The total expenditure was small in relation to the size of the task in hand. By the end of 1938 the Commissioners had committed £17 million, but only one half of this sum had been spent, much having gone on land settlement and public health schemes.[6] Less than £3 million had been spent on assisting the establishment of new private enterprises, involving 121 firms and the creation of 14,900 jobs. In addition, a further 290 firms had been assisted by capital or other inducements, while in the later 1930s a number of foreign firms were enticed to locate in the designated areas. Perhaps the most successful part of the policy was that relating to the creation of trading estates, the forerunners of which were those at Trafford Park, Slough, Welwyn and Letchworth. Government trading estates were established at Treforest, Team Valley and Hillington with minor ones at Pallion (Sunderland), St. Helens Auckland, Larkhall and Cyfarthfa. By the end of the decade a total of some 12,000 workers were employed on these estates.[7]

The overall results of the special areas legislation were relatively meagre. Probably less than 50,000 new jobs were created as against 362,000 registered as unemployed in the special areas at the beginning of 1935. National recovery forces did more than regional policy to reduce unemployment, though even by the end of 1937 unemployment in these areas was still twice the national average. In the initial years the Commissioners faced a difficult task given their limited powers and lack of finance, and even when these were extended later in the 1930s it still proved difficult to entice firms to the depressed areas. Moreover, the regions selected for assistance were few in number and their boundaries were not such as to facilitate long-range planning. The government conceived the policy in terms of a temporary expedient and refused to extend its geograpical coverage, so depressed areas such as Lancashire and

Merseyside were excluded from its scope. For the most part the government preferred to rely on providing the unemployed with maintenance relief rather than with job opportunities. Thus the effort to stimulate employment in these areas may be seen as little more than half-hearted concessions to a growing public concern over the persistence of continuous unemployment.[8]

A case can certainly be made for a stronger regional policy, and in particular one which encouraged the establishment of new growth sectors and the building industries with strong employment multiplier effects.[9] However, having said this, it is apparent that it would not have been a very easy policy to implement, short of state direction of industry, and even then it would have taken many years to have an appreciable impact. Moreover, the one example we have of an induced move, that of Richard Thomas's new steel works to Ebbw Vale, can scarcely be accorded high praise in terms of optimal location.[10] The fact is that the decaying northern regions offered an unattractive prospect for private enterprise location compared with the South and Midlands. In terms of market structure (levels of income and size of population), accessibility, labour supply characteristics, infrastructure facilities, scale economies and general environment, they scored badly *vis-à-vis* the South. In addition, new forms of energy (electricity) and transport (motor transport), by making industry more footloose, reduced the attractions of the North still further, while favouring the light, high value-added assembly industries of the South. The cumulative weight of these factors was sufficient to deter capital and enterprise from moving into the poorer regions.[11] In a survey of 5,800 firms conducted in 1935, Sir Malcolm Stewart (Special Areas Commissioner for England and Wales) found that only eight were even prepared to consider investing in the special areas. The reasons cited for not so doing were inaccessibility, low consumer purchasing power, the high degree of unionisation and heavy local taxes.[12] One might also add that the dead-weight of a traditional industrial structure coupled with the general air of decay tended to stifle change and adaptation.

The difficulties facing the rejuvenation of depressed areas were summed up graphically by Ellen Wilkinson in her study of Jarrow, one of the worst-hit towns of the North:

> When the industry of a town has been killed, it seems as difficult to apply artificial respiration as on a human corpse. This is particularly true when the town or district has been a one-industry one like Jarrow or South-West Durham or the cotton towns of Lancashire. All the

> traditions of the area, all the specialized skill, seem to cling to the dead industry. Though human beings prove themselves very educable and adaptable when they go elsewhere, the assumption is made that in their own town they can only do what they have always done.[13]

Apart from the more obvious disadvantages of the northern regions as locations for new enterprise, the failure of the depressed areas to gain re-entry into the national economy can also be attributed to the type of industrial organisation and structure prevailing there. It lacked the degree of flexibility evident in the south because of the heavy commitment to one or two ex-growth staple industries and the domination of large conservative firms which deterred the development of new enterprise, both indigenous and in-migrating. By contrast, the South and Midlands had a much more diversified and flexible industrial structure which facilitated the entry of new firms in rapidly growing sectors of activity. Carol Heim has recently enlarged upon this theme by suggesting that the reintegration of the depressed areas into the more prosperous parts of the economy was inhibited by organisational constraints. She argues that large firms in the growth sectors had not yet reached the stage when they were able or willing to insert branch factories into the depressed areas as they did after the second world war, though clearly the easier labour supply situation may account for some of the lack of enterprise mobility in the earlier period. Secondly, the barriers to entry in new fields of endeavour became significant in the 1930s because of competition from already established producers in the South which initially had a headstart in the growth sectors, together with the development of controlled competition by restrictive trading practices in the very fields of activity to which it was essential that the depressed regions should gain entry if they were to transform their industrial structure. In other words, the depressed areas were caught between two stools. It was too late for producers to enter many of the expanding industries successfully because of the headstart made elsewhere and the growing restrictions on entry, while at the same time it was too early for large producers outwith these regions to be dispersing their activities through branch plant location in the depressed areas to enable them to re-integrate into the national economy. For these reasons the author feels that the scope for expansionary macroeconomic policy would have been limited: 'Interwar aggregate demand policy alone could not rapidly produce the organisational prerequisites for a new role for the depressed areas.'[14]

In addition to the work of the Area Commissioners, several less formal attempts were made to deal with the problem of the depressed areas. In the mid 1930s the Special Areas Reconstruction Association Ltd. (SARA) was established under the aegis of the Bank of England and with partial Treasury backing, to promote industrial diversification in the depressed areas. The geographical coverage was the same as that of the Area Commissioners and most of the assistance was offered to small firms. This was supplemented by the work of a private organisation, the Nuffield Trust established by a gift of £2 million from Lord Nuffield, and a Treasury Fund of £2 million which was later set up in response to criticisms about the work of SARA. All told the total amount of money disbursed under these experiments in regionalism was very small. The employment created could have been no more than 25,000 at a maximum, and their activities did little to bring about the diversification required.[15] Most of the assistance went to small firms and light industries with limited prospects, rather than on new industries proper, and the size of the funds available was minute in relation to the task in hand.

Overall therefore, regional policy created only a few new jobs in the worst-hit areas of the country, and it certainly did little to transform the prospects of the depressed areas. The limited scope of the policy together with the meagre financial support inevitably meant that the impact would be very modest. The limited appraisal of the regions' needs was in part a reflection of the prevailing practices and beliefs of the financial community (including the Treasury), which guaranteed that the policy measures would be small and ineffective. The financial community was reluctant to see extensive government intervention in the industrial sphere, and was unwilling to be drawn into large-scale schemes of assistance which might upset market forces. The Bank of England for its part was more concerned with rationalising and resurrecting the staple industries and had no clear conception of what a policy of regional industrial diversification involved. As Carol Heim writes: 'Finance might not have been the main cause of, or solution to, Britain's industrial difficulties, but the lack of attention to industrial diversification, as opposed to rationalisation, clearly hurt Britain in subsequent years.'[16]

## Industrial Policy and Rationalisation

If regional policy failed to create many new jobs, the government's industrial policy produced even less. Not that there was a consistent and coherent industrial policy worthy of the name, more a hotch-potch of *ad hoc* measures to meet particular circumstances. Broadly they comprised two forms: (1) various types of financial assistance including subsidies and loan guarantees, and (2) a variety of government-sponsored restriction-*cum*-rationalisation schemes designed primarily to support the old staple industries. Whatever else they may have achieved, their impact on employment was slight, and for obvious reasons. The amount of financial aid dispersed was very small and secondly, insofar as the policies under (2) were designed to eliminate excess capacity and improve efficiency they tended to have an adverse effect on employment.

Although public expenditure on economic services was fairly high in this period, accounting for some 15 per cent of total public outlay in the later 1930s, it was not specifically designed to relieve industrial unemployment in the worst-hit areas. A substantial part of it consisted of capital spending on highways, electricity generation, by the Post Office, on trading services of local authorities and on construction work. Most of the remainder could be classed as current expenditure in aid of industry and agriculture; in 1935 it amounted to nearly £29 million, over two thirds of which went on agriculture.[17] This figure does not include expenditure on interest and loan guarantees or on housing subsidies.

There was no particular pattern about the aid given to industry and agriculture. It tended to fluctuate more violently than public expenditure as a whole and to be cut back sharply in times of crisis, as in 1931-32. There was no planned programme of assistance, merely 'a multiplicity of measures which were small in scale and unco-ordinated.'[18] The principles on which aid was dispersed were never clearly defined and assistance was given to both old and new industries.

The chief forms of financial assistance were subsidies and interest and loan guarantees. In the 1930s the chief beneficiaries of subsidies were housing, shipping, aviation and agriculture. Treasury subsidies to promote both local authority and private enterprise house-building had been extensive in the 1920s, but in the following decade they were confined largely to dealing with the twin problems of slum clearance and overcrowding following legislation passed in 1930 and 1935. Altogether some 300,000 houses were built in England and Wales under this legislation, the majority by local authorities.

The shipping industry had long benefited from concealed subsidies for carrying the mail and during the 1920s shipbuilding secured extensive loan guarantees under the Trade Facilities Acts of 1921-26. During the 1930s however more direct aid was doled out to the maritime industries. Following a request from the Chamber of Shipping for a temporary subsidy for the worst-hit sector of the industry, tramp shipping, the government passed the British Shipping (Assistance) Act in 1935, providing a subsidy of £2 million and introducing a 'scrap and build' scheme. The subsidy provisions proved the more successful of the two measures. The amount of assistance was not to exceed £2 million in any one year and it was to be reduced progressively as freight rates returned to their 1929 level. Moreover, the money was only made available on condition that tramp-shipping owners took action to reduce competition and improve the level of freights. This they did by securing the cooperation of shipowners in all the main tramp-owning countries except Japan, and by establishing minimum freight rate schemes in three major trades, the Australian, the Argentinian and the North Atlantic. By 1937 freight rates had improved sufficiently to allow the subsidy provisions to lapse. In all, tramp-owners received a useful windfall of £4 million which probably helped to prevent the financial collapse of this sector.

The 'scrap and build' scheme was designed to assist both the shipping and shipbuilding industries by encouraging the elimination of redundant or excess tonnage and stimulating owners to modernise their fleets. For each ton of new shipping built, owners were required to scrap two tons and one ton for every ton modernised; in return the Treasury advanced money or guaranteed loans up to a maximum of £10 million which it was estimated would be sufficient to build 600,000 gross tons of modern cargo vessels. Unfortunately, the scheme proved a great disappointment. Shipowners were opposed to it from the start on the grounds that it would prove of greater benefit to the shipbuilders. In actual fact it proved of little benefit to either party. Only 50 vessels of 186,000 gross tons were in fact built, for which £3½ million was advanced out of a permitted total of £10 million; 97 vessels of 356,625 gross tons were nominated for demolition, but no applications were received for modernisation work. It appears that the financial provisions of the Act were not sufficiently attractive to encourage large-scale scrapping of fleets, and as freight rates began to improve from 1935 onwards owners were keen to hang on to their old tonnage in the hope that it could be found employment at a later date. Moreover, as British owners were allowed to acquire foreign vessels for scrapping, the chief beneficiaries of the scheme appear to have been foreign shipowners who sold their old ships to owners in this country at inflated prices. Most of the new building under the scheme went to the North-east coast, Scotland's share being minimal. The poor response to the scheme and impending strategic needs later in the decade prompted the government to introduce a more generous scheme in 1939 under which orders for 700,000 gross tons were received by British yards.[19]

Finally, the Cunard Company received a low-interest loan of £8 million towards the cost of building two crack liners for the transatlantic service which gave a much needed boost to John Brown's depressed shipyard.

The only other major recipients of subsidies were the railways and aviation. In the mid-1930s substantial loans (£32 million) were made available to the railways for schemes of modernisation, mainly electrification. The chief beneficiary was the Southern Railway Company, the least seriously depressed of the four main railway companies. At the same time civil aviation was being offered even larger hand-outs than it had received in the previous decade. Civil aviation had never been able to 'fly by itself' and so from its inception it had been a heavily subsidised operation. Subsidy payments to Imperial Airways, the chief operator, amounted to £0.5 million in 1935, without which it would have made a substantial loss. The annual subsidy limit to aviation was raised to £1½ million in 1936 following the establishment of the government's second 'chosen instrument', British Airways, to develop the European route network. Two years later it was raised again to £3 million and a small subsidy of £100,000 was also made available to the domestic airlines. But for these subsidies it is doubtful whether aviation would have progressed very far since few operators made a clear profit during the period.

The only other financial aid worth mentioning is that relating to exports. Shortly after the war a scheme administered by a department within the Board of Trade was set up to insure exporters' credit risks. It was originally conceived as a temporary expedient for dealing with trading difficulties likely to arise during the Armistice period, but in view of the subsequently depressed state of the export trade it was continued on an *ad hoc* basis and eventually made permanent at the end of the 1930s. Most of its business in fact occurred in this decade, when total guarantees were well over £200 million as against a mere £19 million in the 1920s. Even so, the total sum insured was small in relation to Britain's export trade and it is doubtful whether it did much to help the exports of the depressed staple industries.[20] That apart, not a great deal was done to stimulate exports directly despite the half-hearted attempts to encourage Colonial and Empire trade and development through the Colonial Development Fund and the Empire Marketing Board.

The second main area of government interest in the industrial field was more direct intervention affecting the structure of particular industries. This took several forms, not all of which were mutually consistent, in an attempt to deal with the structural problems of the older sectors of industry. It comprised rationalisation to cut out excess capacity and improve efficiency, reorganisation favouring larger and more economic units, and the establishment of restrictive trading practices such as quotas and price support schemes which tended to defer the removal of excess capacity thereby partly nullifying the first two objectives. In part the new approach can be seen as a continuation of industry's own attempts to meet the changed conditions of the postwar period by amalgamation, capacity elimination and the establishment of restrictive support schemes, which led to a greater degree of concentration and collusion in industry during the interwar period. It also gained the support of the banking community through the Bankers' Industrial Development Company, a concern set up in 1930 by the Bank of England and the clearing banks. Its principal tasks were those of rationalising and reorganising the capacity of the ailing industries and its main activities were centred upon the shipbuilding, shipping, textile and iron and steel industries. Intervention at the micro level also accorded with the growing concern among middle-ground political opinion about the need for some kind of planning and support to facilitate the coordination and rationalisation of industrial activities. One of the leading advocates of micro intervention was Harold Macmillan who developed his own particular brand of industrial planning in a series of studies published in the 1930s. The government's own approach fell rather short of the version advocated by the apostles of industrial planning but it provided a convenient and less unpleasant alternative to a more active fiscal policy.[21]

Apart from the shipping industry, which has been discussed above, three other major industries — coal, iron and steel and cotton — were subject to some form of state intervention in this period. The first to be dealt with was the coal industry under the Coal Mines Act of 1930 which attempted to strengthen the rather weak efforts made by the industry itself during the 1920s to regulate its activites. Under Part I of the legislation compulsory cartelisation schemes run by the owners were to allocate output quotas and fix minimum prices for different grades of coal, while the second Part of the Act created a Reorganisation Commission to promote amalgamations by compulsion if necessary. The results of the legislation were disappointing to say the least, partly because the two Parts of the Act were mutually contradictory. It did little to ensure the long-run viability of the industry or to adjust capacity to the changed demand conditions. Insofar as the marketing schemes were successful — though inter-district price competition still continued to prevail — the tendency was for the available business to be spread among all concerns regardless of their relative efficiency, so that all enjoyed the benefits of fixed prices and restricted output. Moreover, since the provisions of the legislation actually impeded the transfer of production quotas to the most efficient units, the problem of capacity was not tackled seriously. Indeed, the quota system tended to perpetuate the existence of collieries which in normal circumstances would have been eliminated, and as a result the rate of structural change in terms of capacity and output slowed down quite noticeably in the 1930s. The operation of Part II of the Act proved a dead letter. The attempts by the Commission to bring about amalgamations and concentrate production were frustrated by determined opposition on the part of both owners and miners, though for different reasons — loss of independence in the case of the former and the threat of unemployment from the mineworkers' point of view. Hence the experiment in rationalisation proved totally abortive apart from the industry's own voluntary efforts, which were more limited than they had been previously simply because the cartel schemes helped to preserve the *status quo*. The powers of the Reorganisation Commission were eventually transferred to the Coal Commission by the Act of 1938 which nationalised coal royalties.[22]

In the case of the iron and steel industry a rather more informal or indirect approach to rationalisation was adopted. Considerable hope was held out for the iron and steel industry during the 1930s when the Import Duties Advisory Committee recommended stiff tariff protection on condition that reorganisation of the industry would follow. To assist the process the central organisation of the industry was strengthened by the establishment, in April 1934, of the British Iron and Steel Federation, one of whose tasks was to foster schemes of reorganisation. In effect these two bodies became responsible for the joint supervision of the industry with responsibility for price fixing, the control of competition, subsidising high-cost producers and supervising the development plans of the industry. Thus for the first time a degree of central direction was exercised over the investment and production plans of the individual companies, though outright amalgamation and capacity elimination were left to the industry's own initiative. In practice the suppression of competition rather than technical reconstruction and rationalisation of the industry became the primary objective of the controlling authorities. Despite considerable amalgamation activity during the period, the principles of plant specialisation, physical integration of units, the adoption of best practice techniques and optimal plant location were only tentatively acknowledged. As one contemporary observer remarked: 'the lamentable history of the attempt to reorganise the iron and steel industry on its present basis, in return for protection, seems to indicate that a very wide measure of public control will be necessary if the badly needed work of rationalisation is ever to make any real progress.'[23] Whether greater public control would have been any more effective remains an open question since in the one instance of government intervention — that of forcing Richard Thomas and Co. to site their new strip mill at Ebbw Vale — the location was decidedly less optimal than the original one in Lincolnshire.[24]

The experience of the cotton industry differs somewhat from that of coal since despite a sharp contraction in demand during the 1920s it was not until the following decade that there was any significant reduction in capacity. The large number of small independent producers and the high degree of horizontal competition in the various sections of the industry made it extremely difficult to reach formal agreement to rationalise the industry, while the continuing support of the banks also delayed capacity adjustment.[25] The first real attempt to deal with excess capacity did not occur until the end of the 1920s, when the Lancashire Cotton Corporation and the Combined Egyptian Mills Ltd. were set up to reduce spindle capacity; both of these groups secured the backing of the major banks and the Bankers' Industrial Development Corporation. It was not until the mid-1930s however that voluntary action was reinforced by legislation, partly because the disastrous consequences of the reorganisation plans for coal had discouraged the government from introducing a similar plan for cotton despite tentative draft proposals prepared by the Board of Trade in 1931. When intervention finally came it was initially confined to capacity reduction. The Cotton Industry (Reorganisation) Act of 1936 set up a Spindles Board to buy up and scrap 10 million surplus spindles, the scheme to be financed by a compulsory levy on existing machinery. By the beginning of the second world war the Board had managed to sterilise about half the targeted amount. Reductions in capacity were also achieved in other sections of the industry though action was much less formally organised than on the spinning side. Nevertheless, despite the lack of official intervention, the number of looms actually fell by more than that of spindles — 37.7 per cent as against 30.4 per cent between 1929-38.[26] Yet despite the much reduced capacity of the cotton industry, the process of adjustment still had some way to go by 1938 since only two thirds to three-quarters of the industry's capacity was being effectively employed. Ironically, in the following year, the government acceded to the industry's request to introduce price fixing and quota schemes by the Cotton Industry (Reorganisation) Act of 1939, which would presumably have had the same effect of slowing down capacity adjustment as in the case of coal. Fortunately, the outbreak of war prevented the implementation of these propsals in their original form.

On the whole the government's industrial policy, such as it was, did not achieve very much. Here and there it did help to stabilise prices and reduce competition, and on a broader front it did confer official blessing on the trend towards concentration and cartelisation that was taking place throughout industry during this period. But official intervention in industry was too limited in scope and too ill-conceived to make much contribution to recovery or to the elimination of capacity in general and improvement in efficiency. As Kirby has noted: 'under the pressure of rapidly changing conditions ministers were compelled, without reference to a set of guiding principles, to take successive *ad hoc* steps to help those industries and interests which were least able to cope with these conditions and whose immediate plight was the most serious for the country as a whole.'[27] Many of the arrangements were designed to protect the producer at the expense of the consumer and since they tended to protect all producers, whether efficient or not, they reduced the flexibility of the industrial structure and slowed down the scrapping of redundant capacity. There was, moreover, a basic conflict of interest between employment generation and capacity elimination, the most notable instance being the legislation dealing with the coal industry. It was perhaps unfortunate that the industries most favoured with intervention and protection were the older ones with substantial excess capacity since these required effective rationalisation to ensure their long-term viability. This could be secured either by competition or through organised control but not by legislation with mutually incompatible objectives.

## Agricultural Policy

In the 1930s agriculture became one of the most highly protected, heavily subsidised and organised of all British industries. This special treatment followed a period in which agriculture had been left very much to fend for itself. Soon after the wartime profits bonanza, falling world prices, increasing competition and withdrawal of government support schemes hit the industry hard.[28] The natural tendency therefore was for the industry to swing back to its prewar pattern, that is away from cereal crops and to concentrate on fruit, dairy produce and meat production. This added to the already increasing number of unemployed agricultural workers since pasture requires from seven to nine less men per thousand acres than arable land. However, despite rising home demand for these products, falling costs abroad and more efficient methods of packing, grading and preserving produce meant that the domestic industry faced fierce foreign competition. Faced with these burdens agriculture became a declining and generally depressed industry. Only sporadic attempts were made by the government to relieve the industry's misery, including the encouragement of smallholdings by Acts of 1919, 1926 and 1931, improved credit facilities (1928), the derating of land and buildings (1929) and the subsidisation of beet sugar production (1925). The last of these proved a costly exercise though it did at least provide employment for some 32,000 workers in agriculture, apart from those engaged in several factories which were amalgamated into the British Sugar Corporation in 1936.

Such measures provided only marginal relief to a hard-pressed industry and they could do little to shield a high-cost producer from the onslaught of the massive fall in primary product prices during the great depression. The National Government therefore, conscious of the social and strategic importance of a prosperous agriculture, launched an extensive programme of protection and assistance, the main legislative basis for which was the Import Duties Act of 1932, the Wheat Act of the same year, and the Agricultural Marketing Acts of 1931 and 1933. Briefly, these provided the industry with extensive protection, subsidies for various products, and producer marketing boards.

The most direct subsidy, apart from that for sugar beet production already in existence, was the one for wheat under the Act of 1932. Producers were guaranteed a standard price of 10s per cwt., below which the government made up the differential by a subsidy, though with a ceiling on output to discourage excessive production in an already overstocked wheat market. Subsidies were also paid at various times to the livestock industry and to producers of oats and barley, while the beet sugar subsidy was continued indefinitely.

As far as protection was concerned, the position was complicated initially by the fact that the general tariff of 1932 and the subsequent

Ottawa agreements left agriculture rather exposed to competition from Dominion producers who, under the system of Imperial preference, enjoyed almost open access to the British market. The problem was solved by imposing fairly severe restrictions on foreign imports, together with some quota limitations on food imports from the Empire. In 1933 and 1934 quota restrictions were imposed on imports of meat, bacon, dairy products, eggs, poultry, potatoes and sea fish. Thus in the case of bacon, Danish imports were severely restricted by voluntary agreement and compulsory quotas, with the result that they fell from 12 to 7½ million cwt between 1929-38, over one quarter of which was made good by imports from Canada. British pig producers benefited considerably, the number of pigs increasing by nearly 20 per cent during this period.

The third measure of assistance comprised the setting up of marketing boards for various products, provided two thirds of the producers assented. The stated intention of the boards was that of rationalising marketing methods, effectively a euphemism for controlling output and prices, for which the government could throw in additional protection if so desired. Marketing boards were subsequently set up for milk, potatoes, pigs, bacon and hops along these principles, the most successful of which was that for potatoes and the least effective those for bacon and pigs.

Agriculture could scarcely claim that it was neglected in the 1930s, though equally it could argue that extensive intervention did not restore it to its former glory. Output barely changed in the 1930s, remaining below the prewar level, while prices were still down on those of 1929. Agriculture's relative size in the national economy continued to decline in terms of both output and employment, while migration from the land continued steadily; total employment, excluding farmers the numbers of whom remained fairly static at around 225,000, declined from 742,000 in 1930 to 593,000 in 1938. The loss in manpower was offset by a steady rise in productivity due to the spread of mechanisation and the adoption of new techniques and management methods. By the end of the decade livestock products accounted for some 70-75 per cent of the value of agricultural output, with farm crops and horticultural products accounting for 16.1 and 13.4 per cent respectively. Britain maintained a fair degree of self-sufficiency in several products including liquid milk, potatoes, pig-meat, vegetables, poultry and eggs but still depended heavily on foreign supplies for beef, mutton, lamb, cereals, fruits, flour and sugar.

State intervention in agriculture has been criticised on the grounds of cost and efficiency. Domestic food prices were maintained at a higher level than those ruling on the world market and consumers as taxpayers also bore the cost of the heavy annual subsidy bill of around £40 million. Protection and marketing schemes sheltered both efficient and inefficient producers and thereby slowed down the elimination of high-cost production. As Walworth noted at the time, 'British Marketing policy is economically unsound, starting from the basis of securing costs of production plus profit on existing methods of production and covering marginal men.'[29] Possibly however this was a price worth paying in view of the nation's food requirements during the second world war.

## Policy Measures in Retrospect

Despite the significant shift in policy direction and the wide variety of measures taken in the early 1930s, there has been a tendency to play down the role of policy in the recovery of the 1930s. In fact during the course of the above discussion the present writer has probably done the same thing. On further reflection however one should perhaps consider the verdict more closely. We do not doubt that much of the recovery can be attributed to spontaneous causes or non-policy factors, but at the same time it may be wrong to dismiss the latter on the grounds of their marginal importance. After all, most of the policy measures adopted in this period pointed in the right direction, albeit not always very strongly, and it may be argued that in each case a policy-on situation was better for employment than the alternative of no policy change at all. Moreover, one should consider the counterfactual world more closely. Competitive exchange depreciation and trade restrictions may lead to income and trade destruction for the world as a whole, but one country does not gain by standing out against the pack. If Britain had not abandoned gold nor adopted tariff protection in a world which did, she would no doubt have been worse off. The example of France (or other gold bloc countries for that matter) who refused to devalue until the mid-1930s is instructive, since in order to retain an overvalued exchange rate she was forced to compress domestic costs and prices with the result that output and exports continued to stagnate for much of the decade. By devaluation and protection Britain was at least able to provide scope for lower interest rates and domestic expansion.

Some of the policy measures were of course quite helpful, notably cheap money, though it would be difficult to estimate precisely its impact in terms of employment and output. It appears to have been most influential in the housing market, and least effective in terms of stimulating industrial investment, though both these points require modification, as we have seen. Other measures, such as regional and industrial policies, were less impressive, though even the latter could be said to have made a small contribution insofar as they helped to maintain or raise prices since this helped to restore confidence to industrialists whose cost-price relationships had been distorted in the depression (see Chapter 6). Even fiscal policy, which proved to be the biggest failure of all, had something to offer indirectly in as much as the pursuit of sound financial policies helped to restore confidence and pave the way for lower interest rates. Finally, insofar as transfer payments were maintained at a relatively high level, despite the cuts at the bottom of the recession, a floor was set to consumption which helped to ensure an early upturn from the recession. Thus, while policy factors were not the main driving force in the recovery, it is perhaps harsh judgment to write them off altogether.

## Notes

1. E.D. McCullum, 'The Problem of the Depressed Areas in Great Britain', *International Labour Review*, 30 (August 1934), p.137.
2. Political and Economic Planning, *Report on the Location of Industry in Britain* (1939), p.44.
3. C.M.Law, *British Regional Development since World War I* (1980), pp.58, 60, 67.
4. D.E. Pitfield, 'The Quest for an Effective Regional Policy, 1934-37', *Regional Studies*, 12 (1978).
5. J.A. Garraty, *Unemployment in History* (1978), p.211.
6. A. Booth, 'The Second World War and the Origins of Modern Regional Policy', *Economy and Society*, 11 (1982), pp.4-6; J.D. McCallum, 'The Development of Regional Policy', in D. Maclennan and J.B. Parr (eds.), *Regional Policy: Past Experience and New Directions* (1979), p.5.
7. *Report of the Royal Commission on the Distribution of Industrial Population*, Cmd 6153 (1940), pp. 147-8.
8. E.M. Burns, *British Unemployment Programs, 1920-1938* (1941), pp.328-9.
9. See M.E.F. Thomas, 'Regional Unemployment and Policy in the 1930s: A Preliminary Study', *Essex Economic Papers* (1981) and T.J. Hatton, 'Structural Aspects of Unemployment between the Wars', *Essex Economic Papers* (1982).
10. D.E. Pitfield, 'Regional Economic Policy and the Long Run: Innovation and Location in the Iron and Steel Industry', *Business History*, 16 (1974).
11. D.H. Aldcroft, *East Midlands Economy* (1979).
12. F.M. Miller, 'The Unemployment Policy of the National Government, 1931-1936', *Historical Journal*, 19 (1976), p.469.

13. E. Wilkinson, *The Town That Was Murdered* (1939), p.263.
14. C.E. Heim, 'Industrial Organisation and Regional Development in Interwar Britain', *Journal of Economic History*, 43 (1983), pp.931, 934, 942, 947, 950.
15. C.E. Heim, 'Limits to Intervention: the Bank of England and Industrial Diversification in the Depressed Areas', *Economic History Review*, 37 (1984), p.533.
16. *Ibid.*, p.545.
17. A.T. Peacock and J. Wiseman, *The Growth of Public Expenditure in the United Kingdom* (2nd edn., 1967), p.178.
18. W. Ashworth, *An Economic History of England, 1870-1939* (1960), p.405.
19. S.G. Sturmey, *British Shipping and World Competition* (1962), pp.107-10; N.K. Buxton, 'The Scottish Shipbuilding Industry between the Wars: A Comparative Study', *Business History*, 10 (1968), pp.115-16.
20. For details see D.H. Aldcroft, 'The Early History and Development of Export Credit Insurance in Britain, 1919-1939', *The Manchester School*, 30 (1962).
21. See A. Marwick, 'Middle Opinion in the Thirties: Planning, Progress and Political Agreement', *English Historical Review*, 79 (1964) and H. Macmillan, *Winds of Change, 1914-1939* (1966), Ch. 12.
22. M.W. Kirby, 'Government Intervention in Industrial Organisation: Coal Mining in the Nineteen-Thirties', *Business History*, 15 (1973) and 'The Control of Competition in the British Coal-mining Industry in the Thirties', *Economic History Review*, 26 (1973).
23. A.F. Lucas, *Industrial Reconstruction and the Control of Competition* (1937), p.121.
24. D.E. Pitfield, 'Regional Economic Policy and the Long Run: Innovation and Location in the Iron and Steel Industry', *Business History*, 16 (1974).
25. J.H. Porter, 'The Commercial Banks and the Financial Problems of the English Cotton Industry 1919-1939', *International Review of Banking History*, 9 (1974) and W. Lazonick, 'Competition, Specialisation and Industrial Decline', *Journal of Economic History*, 41 (1981) and 'Industrial Organisation and Technical Change: The Decline of the British Cotton Industry', *Business History Review*, (1982).
26. M.W. Kirby, 'The Lancashire Cotton Industry in the Inter-war Years: A Study in Organisation and Change', *Business History*, 16 (1974).
27. M.W. Kirby, *The Decline of British Economic Power since 1870* (1981), pp.67-8.
28. P.E. Dewey, 'British Farming Profits and Government Policy during the First World War', *Economic History Review*, 37 (1984).
29. G. Walworth, *Feeding the Nation in Peace and War* (1940), p.503; E.H. Whetham, *The Agrarian History of England and Wales*, Vol. 8, *1914-39* (1978).

# 6 Natural Forces of Recovery

The debate about Britain's strong recovery in the 1930s has centred on two main issues. The first of these relates to the question of whether or not it can be regarded as a spontaneous recovery, that is one induced largely by natural or real factors as opposed to a policy dominated one. The second and more controversial issue is concerned with the relative contribution to recovery of the various natural forces. The bulk of this chapter is concerned with the second issue, but first a comment on the question of spontaneity.

## How Spontaneous was the Recovery

The first problem need not detain us long since we have already looked in detail at the contribution of policy factors. We know for certain that much of the increased activity which occurred in the 1930s was on account of the private sector; even in housebuilding, where the state had played a major role during the 1920s, the bulk of the new dwellings constructed in the 1930s was provided by private enterprise. We also know that there was no significant rise in public spending and public investment in the early years of recovery so that most of the new employment generated came from the private sector. Deficit spending was not pressed into service to inflate the economy and fiscal policy acted as a drag on the economy during the early years of recovery. On the other hand, by the later 1930s, as a result of the rearmament programme, spending increased sharply and, as we have seen, it helped to moderate the downturn of 1937-38 and, according to recent estimates, provided a significant boost to employment in some of the basic sectors of the economy. Nevertheless, while direct government intervention in terms of public spending made no contribution to recovery in the initial stages, several important policy changes indirectly assisted the recovery process. Cheap money, for example, certainly aided the housebuilding sector even though it cannot be regarded as vital to the vigorous housebuilding boom, but its influence on non-residential investment was weak. Devaluation and tariff protection clearly did not spark off an export-led recovery but they did help to stabilise the British trade position and eased the constraint on the external account which then made possible a sustained rise in domestic output. Indeed, it is possible to argue that had the authorities not released sterling in 1931, the British economy would have gone the same way as the French economy in the 1930s as a result of the internal compression required to maintain the former parity of the exchange. Moreover, these measures, together with the government's tight fiscal stance, helped to pave the way for cheap money, which regardless of its merits and influence, was obviously more beneficial than the dear-money policy of the previous decade. Thus while a direct policy stimulus was conspicuous by its absence, the indirect benefits flowing from the several policy changes of the early 1930s may be of more substance than many writers have been prepared to acknowledge. A recent contribution to the literature by Worswick concludes that spontaneous forces alone are not enough to account for the recovery and that policy made important contributions.[1]

Nevertheless, most commentators on the period have acknowledged that there was a strong degree of spontaneity in the recovery, which has given rise to a heated debate about the relative importance of different natural forces. The controversy has revolved around three major issues. The first is concerned with the scope for investment opportunities and the related question of structural change and sector analysis. Secondly, there is the question of the role of the real income effect in generating recovery. Finally, an old debate recently revived, namely the real wage level and profitability, is worth examining. Each of these points merits detailed consideration.

## Investment Opportunities and Structural Change

At a general level there would seem to be a good case for arguing that real recovery forces were strong in the 1930s because of the known range of investment opportunities awaiting exploitation. To all intents and purposes, there should have been an accumulated backlog of investment projects because of the relatively weak nature of the boom in the later 1920s. Unlike the situation in America, where there was a vigorous boom in residential construction and consumer durables, Britain experienced a very muted response in both sectors and therefore investment outlets had not been exhausted by market saturation as was the case in the United States. The weak recovery in that country in the 1930s would appear to confirm this point. Moreover, the belated development of the newer industries in Britain, partly because of low real income growth during the preceding two decades, together with the shortage of housing arising from the wartime backlog, the rise in the rate of family formation and the impediments to private building (for example rent controls), meant that there was plenty of scope for vigorous growth in the 1930s despite the drying up of investment opportunities in many of the old staple industries. The service trades also offered opportunities for expansion since in many cases market penetration was well below the levels reached in the United States. In short, therefore, the way was clear for a strong upswing in several important sectors once the immediate crisis had passed. That it occurred early and strongly may be attributed to three factors. First, the relative mildness of the depression in Britain, and more particularly the fact that business confidence was not shattered, as in the United States and Germany, by the collapse of financial institutions. Any loss of confidence which did occur was partly restored by the reassuring policy measures taken in 1931-32. Secondly, the favourable real income trend in the early 1930s which ensured an adequate level of effective demand (see below). And finally, the policy actions taken to protect the external account which made it possible for domestic expansion to take place without putting undue strain on the balance of payments in the early years of recovery.

The real difficulty begins when we start to disaggregate the economy, because it is here that controversy arises as to which sectors made the most contribution to recovery. Two sectors in particular have been singled out for special attention — building and the new industries — leading sectors thought to have powered the recovery.

In contrast to the heated debate about the role of the newer industries there has been less dispute about the contribution of the building sector. This is especially true for the early years of recovery when output and employment in this industry expanded at a much faster rate than the general average. Building and contracting output rose by more than a third between 1932-35 as against one quarter for all industrial production and 14 per cent for GDP, while the output of the building material trades rose by 43 per cent. The increase is even more dramatic in the case of residential construction; the number of houses built rose by 56 per cent between 1932-35 and in the first two years of recovery the increase in housebuilding accounted for 17 per cent of the rise in GNP. The majority of this new housing was unsubsidised and built by private enterprise. Even more welcome was the fact that building was a labour-intensive occupation and therefore generated many new jobs. The rise in building employment accounted for some 20 per cent of the increase in insured employment between 1932-35, and possibly for as much as 30 per cent if building materials are included, while over the recovery period as a whole, 1932-37, it was equivalent to 13 per cent.[2] In fact, building and related trades increased their share of total industrial employment from 10.4 per cent in 1920 to 15.2 per cent by 1938. Much of the increase in the share of income devoted to investment can be attributed to this sector. By the 1930s nearly one half the total gross domestic investment in the economy was accounted for by building work, as against 44 per cent in the 1920s and barely a third in the period 1900-9, which is equivalent to around 5 per cent of the national income compared with 2-3 per cent pre-war.[3] The increasing size of this sector was largely the consequence of the rapid growth in residential construction and when this peaked in the mid-1930s building lost some of its dynamic characteristics even though non-residential building began to pick up. Investment in dwellings proved remarkably stable during the depression itself when total investment declined by 14 per cent, while in the first two years of recovery fixed investment in dwellings rose by 47 per cent and accounted for nearly 61 per cent of the rise in total capital formation between 1932-34.

Apart from its direct quantitative impact, the expansion in building had two other important advantages. The geographical distribution of building activity was fairly widespread and its regional multiplier effects are known to be strong.[4] Most regions experienced a fairly rapid growth in building employment and some of the depressed areas, notably Northumberland and Durham, faired better in this respect than areas further south. Secondly, and perhaps the most important feature, is the wide repercussions the industry had on the rest of the economy. Many other industries might well have expanded at a comparable rate but their effects on the economy as a whole would have been very much less. The sheer size of the building sector and its nation-wide ramifications were very considerable. Like the vehicle industry, construction is an assembly process which requires a large and varied supply of components or materials from many industries. These include not only the more obvious building materials such as bricks, cement, tiles, glass and timber, but also a host of other products including iron and steel and metals, paints, pipes, stoves, window frames, heating and ventilating apparatus, rubber, cables and electrical wiring and wallpaper. It is significant that employment in all these trades expanded much faster than the average for all industry in the 1930s. An input-output analysis based on the 1935 Census of Production provides some indication of the widespread demand effects for the products of other industries. Altogether, the building industry bought products from 24 of the 36 main industrial groups classified by Barna, with only two industries, food processing and the distributive trades, having a wider range and larger volume of transactions with other industries.[5]

In short, building had a greater impact on the rest of the economy than practically any other industry. Moreover, if we take account of some of the more indirect and less quantifiable effects of increased construction activity, then the importance of the sector is enhanced even further. The vast new housing programme of these years generated a demand for a wide range of fixtures and fittings — linoleum, carpets, curtains, cupboards, crockery and bedding — and for new consumer durables most of which were based on electricity. Virtually all new houses were wired for electricity, and so too were many old ones, with the result that by the end of the period two out of three houses had electric power laid on as against only one in 17 in 1920. The number of electricity consumers jumped dramatically from less than a million in 1920 to 2.844 million in 1929 and to no less than 9 million at the end of the 1930s. Finally, since much of the new residential building took place in the suburbs it stimulated a demand for new forms of transport and additional infrastructure facilities.[6]

While the quantitative importance of building activity in the 1930s is not in doubt, from the technical point of view the industry had little to offer. Methods of construction were very much the same as they had been in the nineteenth century, that is, a vast assembly operation of a large number of different materials by labour-intensive methods and with the aid of a limited amount of capital equipment. There were a number of new developments, including the introduction of new materials and the use of mechanical aids such as steam shovels, cranes and excavators, though most of the latter were only to be found on large building sites. The vast majority of small building firms departed little from traditional practice. There was no massive application of new technology or new processes to revolutionise building operations and hence the industry did not become a centre for the transmission of new techniques to other sectors of the economy. In this respect therefore its importance was much less than that of some of the new industries, for example motor manufacturing and electrical production, both of which had important implications in terms of technical developments and methods of production for the economy at large. In view of the limited technical progress it is not surprising that productivity growth in the building industry was relatively low compared with that in other industries.

Natural forces were important in generating the surge in residential building though not exclusively so since cheap money had a not insignificant part to play. But, as noted in Chapter 4, there is some difficulty with the timing sequence which precludes laying too much stress on cheap money as an instigator of the housing boom.[7] In any case several powerful non-policy factors favoured an uplift in housing in the 1930s. These include the fall in construction costs, rising real incomes, an increase in the rate of family formation, the shortage of housing, changing consumer preferences and the trek to the suburbs. The convergence of so many favourable factors at one point in time would probably have been sufficient to create a housing boom of some magnitude irrespective of any policy inducements.[8]

We now turn to the more difficult structural issue, namely the role of the new industries in the recovery of the 1930s. That British industry was undergoing a process of structural change in the interwar years is beyond dispute. It was part and parcel of a wider European phenomenon of industrial transformation whereby old sectors were giving way to new, the limits to the speed of which were set by the inflexibility of the existing pattern of development and the rate at which real incomes increased.[9] Before 1914 Britain lagged behind her major competitors in several new fields of activity — electrical engineering, motor manufacturing and new chemical compounds — while at the same time she had a proportionately larger share of resources locked up in the old staple trades of textiles, shipbuilding, mining and mechanical engineering, where growth prospects were on the point of expiring. Thus when the crunch came after the first world war with the collapse of export markets for staple products, the new sectors of activity were as yet too small to cope with the problem of the massive redundant resources of the staple trades.

However, the initial lag in the development of new sectors meant that the potential for advancement in the interwar period was that much the greater. Not surprisingly, therefore, these industries became an increasingly important influence on the growth of the economy. They expanded at a faster rate than the national average and accounted for an increasing share of industrial output and employment. Many of the new sectors witnessed a constant stream of innovations and under the influence of rapid technical progress and scale economies, productivity advanced strongly and selling prices were reduced. Some of the more outstanding changes occurred in the electrical and automobile industries where techniques and methods of production and organisation were transformed out of all recognition to those prevailing before 1914. The dramatic changes are well illustrated by the transformation of motor car manufacturing. Before the war the motor car was very much an expensive hand-made product designed for the rich. A large number of firms produced an equally large number of bespoke models, the technical performance of which left much to be desired. In 1913 the total output of vehicles was 34,000, a mere fraction of the American and less than that of France. The great change in structure and production methods occurred after the war when Morris and Austin, benefiting from wartime experience and the example of Ford, introduced the small, mass-produced car to the British market. By the end of the decade these two firms accounted for 60 per cent of British car production, and the total number of firms in the industry had fallen from 90 to 41 (1920-29). Further concentration took place in the 1930s so that by the end of that decade there were only 33 producers, six of which accounted for nearly 90 per cent of the market, while three (Nuffield, Austin and Ford) accounted for about two thirds. Along with this concentration in production there was a marked rise in the output of each firm, a reduction in the number of models produced and a sharp fall in the price of each unit. Thus from a small output of 73,000 motor vehicles in 1922, there was almost continuous expansion to 239,000 in 1929 and 511,000 in 1937. The average factory value of private cars fell from £308 in 1912 to £259 in 1924, £206 in 1930 and £130 in 1935-36, while the average retail price of cars dropped by about 50 per cent between 1924 and the middle of the 1930s. Together with the fall in running costs and improved tehnical performance of the product, this was an important factor in the large expansion of the market during the period. By 1939 there were some 3 million vehicles of all types on Britain's roads, nearly 2 million of which were private cars.

The transformation of the motor industry may have been one of the more spectacular instances of industrial change during the interwar years, but even so similar trends were apparent in most other new industries — electricity generation, electrical manufacturing, rayon and chemicals. Moreover, many of the new sectors had important repercussions on other branches of activity because of their strong inter-industry linkages. Again motor manufacturing is a prime example of the demand generated for a wide range of products for what was basically an assembly operation of a vast number of components and parts supplied from outwith the industry. Thus motor manufacturing stimulated or brought into being a whole range of industries including oil refining, rubber, electrical goods, glass, leather, fabrics, metallurgy and mechanical engineering. In a wider context motor transport encouraged the growth of the suburbs and new infrastructure development. Similarly, the rapid spread in the use of electricity after the establishment of the National Grid led to an increasing demand for electrical products and electrical equipment from industry, private consumers and public institutions. As the price per unit of current fell sharply — by more than one half between the early 1920s and 1938 — so the number of consumers rose dramatically, to reach nearly nine million in 1938.

The big question at this juncture is how important were these new developments to the recovery effort of the 1930s? Some years ago Richardson suggested that *in toto* the newer industries formed an interdependent development block which was instrumental in pulling the economy out of the recession.[10] While he acknowledged the contribution of housing, this was not considered to be the decisive factor; rather it was the vast expansion in the products of the newer industries, under the influence of supply and demand factors, that determined the pace and the strength of the recovery. In particular, the supply side was characterised by an intensified shift of resources from old to new sectors, which benefited from new technologies, new production methods and scale economies. Thus high productivity growth led to falling product prices and an expansion of the market, the latter being bolstered by rising real incomes and shifts in consumer spending patterns and tastes. The thrust of the argument is one of spontaneous recovery due to natural growth forces with government policy playing a minor role.

The thesis is at first sight an attractive one but since it was first published it has come in for considerable criticism. The substance of the attack is discursive however and therefore requires looking at in some detail. The first problem is one of definition. The distinction between old and new sectors is a useful one though by no means always a very clear or meaningful concept. The definition of new industries tends to vary according to each author's preference but often excludes certain important sectors such as electricity generation. Secondly, the new and old concept is not very helpful when applied to certain industries, for example chemicals and engineering, since they comprised both types. Moreover, as Dowie has pointed out, across the whole spectrum of activities making up the industrial production index there was a wide variety of performance in terms of output, productivity and employment growth. Most new industries expanded their output growth rapidly but not all enjoyed above average productivity gains, e.g. electrical manufacturing. Similarly, some older sectors stagnated in terms of output and employment but did quite well in respect of productivity growth. Furthermore, while all new industries were growing not all expanding industries were new ones.[11] Perhaps therefore, as Pollard suggests, 'one is on safer ground with the assertion ... that there were some industries which grew faster than the rest, while others stagnated and even declined, and that it was the former which carried the expansionary drive of the British economy.'[12] This would carry the implication that the basis of economic recovery was rather more widely dispersed than was once believed to be the case.[13]

A more severe attack has been made on the structural issue, that is the timing, extent and strength of the resource shift to the new sectors. The extent of the structural change has been described as modest because the demand for the products of the newer industries lacked social depth.[14] A more detailed analysis by Buxton casts doubt upon their dynamic power in recovery on the grounds that resource transfer was retarded during the slump[15], and that in the ensuing upswing it occurred at a rate no faster than in the 1920s. In other words, he concludes that these industries were not sufficiently predominant to have influenced materially either the extent or the speed of the recovery: 'it seems unlikely that the "new" industries ever became sufficiently dominant during the 1930s to have played a decisive part in the upswing of the economy. Indeed, available evidence suggests that they cannot have played much more than a supporting role to the much stronger influences that were at work.'[16]

In the light of the evidence presented Buxton's conclusions on the role of the new industries in recovery appear somewhat gratuitous. While one may agree that the resource shift was limited during the depression, it is evident on his own admission that by the mid-1930s the new industries were accounting for a significant share of net industrial output. In 1935 they accounted for 21 per cent of net output as against 15.9 per cent in 1930 and 14.1 per cent in 1924; during the same period the share of the staple trades fell from 37 to 27.8 per cent (1924-35). This would seem to be a not inconsiderable transfer, and by the middle of the 1930s the new industries (which, incidentally, as defined do not include aircraft manufacturing, precision instruments and electricity generation)[17] can scarcely be regarded as insignificant in terms of industrial output at least. Moreover, as he readily acknowledges, the staple trades as a group continued to dis-invest throughout the years 1932-37 (though at a reduced rate from the previous decade), while employment in these industries, though rising gradually from the 1932 trough, never managed to regain the pre-depression levels, averaging about 85 per cent of the 1924-29 level during the course of the upswing. Nor is it clear from Buxton's analysis what the main driving force in recovery was, unless it is building which he sees as the main beneficiary of structural change at the expense of the service and staple trades.[18]

More recently, Broadberry has mounted a broader attack on the concept of industrial regeneration and structural change. His main argument is that throughout the interwar years the economy was in a state of disequilibrium that was not self-adjusting because of a deficiency in aggregate demand. The declining sectors contracted too rapidly with little to replace them so that the quantity signals of reduced demand from the declining sectors swamped the relative price signals and thereby dampened investment and activity in the whole economy. In effect this amounts to a rejection of what he calls the 'Old Optimists' thesis (among whom he includes Richardson and the present author) of industrial regeneration and structural change, for which he gives the following reasons: (1) that productivity growth was not particularly rapid if account is taken of the qualitative changes in the labour force; (2) what productivity growth there was can be explained in part by the elimination of the least efficient firms; (3) structural changes were not especially rapid during this period; (4) the importance of the new industries has been exaggerated; (5) rationalisation does not appear to have assisted productivity growth to any significant extent.[19]

Some of the assumptions can be questioned, others have been dealt with elsewhere. In defence, it should be pointed out that the 'Old Optimists' did not claim that structural change or industrial regeneration provided a complete solution to the disequilibrium problem, nor that progress in this regard was optimal. Clearly they did not, but it would be fanciful to argue that structural changes in this period were minimal given the known facts. The new industries were growing faster than the rest of manufacturing throughout the interwar years and on any definition they were accounting for an increasing share of net output and employment. The increase in the contribution of these sectors to growth came about largely because of their greater weight in the economy rather than by virtue of the fact that they were outstripping other industries to a greater degree than formerly. Moreover, as Matthews and his co-authors suggest, the indirect effects of their growth on other sectors of the economy in terms of output and productivity may have been important in some cases such as electricity and electrical engineering by virtue of the fact that they produced mainly capital goods.[20] Had there been no industrial transformation after the first world war, the prospects for interwar growth and recovery in the 1930s would have been very bleak indeed given the sharp collapse of the traditional export trades. Quite how the process of contraction in the latter sectors could have been modified in the face of severe external shocks is not made fully clear. The author alludes to the need for greater government intervention, presumably to boost aggregate demand, but it is doubtful whether this would have done much to ease the plight of these industries or solved the structural problem in general, given their adverse demand schedules and their obvious supply limitations (inelastic demand and high-cost production). Indeed, such an approach might have helped to ossify the existing structural format of the economy. As Glynn and Booth rightly point out: 'the simple Keynesian approach to interwar unemployment has accorded insufficient attention to this essentially structural and regional problem.'[21]

One final criticism of the new industries again relates to an aspect of the structural question. In a study of structural change in British manufacturing industry between 1907-68, von Tunzelmann seeks to downgrade the traditional overcommitment view of the deadweight of manpower and resources in the declining export staples. If anything, he argues, their shedding of labour accounted for more of the rise in productivity for the country as a whole than did the rise in the share of labour in fast growing industries. Thus in the long run the 'shift' effect of resource transfer among sectors has been much less important in explaining productivity growth than the 'internal' effect of productivity growth within all sectors. Such a shift as occurred was limited to that of a few low-productivity staples and the recruitment of labour into one or two new industries, whereas the internal effect was much more pervasive in both old and new industries. Moreover, apart from a few notable exceptions, for example rubber and non-ferrous metals, the backward and forward linkages of the leading sectors were consistently to the export staples. In other words, it is misleading to view the new industries as a 'development block' largely independent of the staple trades. Nor can the structural shift towards the new sectors fully explain the simultaneous presence of high economy-wide growth rates and heavy unemployment in the 1930s. Instead, 'the advantage of new industries to the economy emerges not so much as exceptionally rapid technical progress (unless through leading innovations elsewhere) nor as once-for-all productivity gains while they are growing rapidly, but equally as the simple Keynesian effect of employing primary resources in sufficiently large quantities.'[22]

This last point perhaps provides a convenient way of salvaging a compromise from the wreckage of this continuing dispute. Neither building nor the new industries were sufficiently robust to bear the full weight of recovery alone, while the staple industries, for obvious reasons, and the service trades, whose employment growth was steady but output growth unspectacular, cannot be pressed into service for this purpose. On the other hand, building and the new industries together provide an admirable combination, with sufficient size and strength to exert a powerful leverage on the economy. Though Richardson highlighted the importance of the new industries, his original conception was of a rather broader-based recovery located in the interaction of these two large sectors. And with good reason. Both responded early and strongly to the upturn and their expansion was sustained for much of the decade. Secondly, they accounted for a large and increasing share of national resources. The combined labour force of the two sectors amounted to 21.3 per cent of the industrial labour force (including mining) in 1920, 27.9 per cent in 1929 and 33.4 per cent in 1938[23], while they accounted for well over one half of total gross investment in the 1930s.[24] Thirdly, and perhaps most important of all, these sectors exerted a considerable impact on the rest of the economy. The new industries and the building trades not only generated a demand for each other's products, but they also had very strong forward and backward linkages on the rest of the economy including some of the staple trades such as iron and steel, mechanical engineering, the metal trades and certain branches of textiles. Rapid growth in these sectors therefore helped to pull the rest of the economy out of recession, which goes too far to explain the vigorous recovery in iron and steel and the metal trades in the first half of the 1930s, and also in many smaller branches of activity. As we have seen, the building trades and new industries required supplies and components from a wide range of industries and so it would be surprising had they not exerted a favourable impact on many other sectors of the economy. Moreover, the building of so many new houses in the 1930s (some 3 million) had an impact far beyond the immediate range of the industries supplying the construction industry. New housing developments stimulated a demand for a wide range of new products for furnishing the home and indirectly created a need for many new infrastructure facilities such as schools, shops, banks, churches, entertainment facilities, restaurants, garages, public utilities and new transport facilities, and since most houses were being wired for electricity this also helped indirectly to generate a demand for a wide range of new consumer durables based on the new form of power.

In other words, the strength of these two large sectors not only determined the speed and vigour of the recovery but also ensured that it would become a broadly-based one as their repercussions spread far and wide throughout the economy. Even in the early years of recovery, 1932-35, few industries failed to expand their output and employment, though it is worth noting that the strongest response occurred in the building trades, new industries, and in the engineering, metal and iron and steel industries. The main laggards comprised agriculture, mining, textiles, clothing and transport (excluding road transport). In the later 1930s most of the new industries maintained their momentum, the building trades peaked out, while some of the heavy metal industries were boosted by the beginnings of rearmament. If we look at through cyclical performance (1929-37) then the contrast between the leading sectors and the old staples is even more pronounced. Agriculture, mining, textiles, shipbuilding and iron and steel are found to be stagnating or declining areas in either output or employment or both, while mechanical engineering, clothing, drink, and transport put up a rather desultory performance compared with the rest of the economy. By contrast, the big growth sectors in terms of both output and employment were building and allied trades, chemicals, electrical engineering, vehicle manufacture, paper, printing and publishing, the metal trades, gas, water and electricity and food manufacture. The service trades also show a strong response in terms of employment with less fluctuation compared with the industrial sector, but their contribution to output and productivity growth was very modest.[25]

The structural pattern of Britain's recovery in the 1930s may be summarised as follows. It was primarily a home-market recovery triggered off by strong natural growth forces in key leading sectors, building and allied trades and the new industries, the impact of which quickly spread to many other sectors of the economy so that recovery soon became broadly based. When construction began to wane in the middle of the decade the recovery momentum elsewhere was strong enough to sustain the process until rearmament gave a much needed boost to some of the old staples. Underpinning the growth and transformation of these years were the real income gains of the early 1930s to which we now turn.

## Real Incomes and Effective Demand

The second major factor responsible for the strength of the recovery was the favourable real income trends and the high floor to consumption through the depression. Despite the ever-present spectre of poverty and the alleged deficiency of demand, the fact remains that for those in work the interwar period as a whole was a time of rising living standards. Real national income rose by about 30 per cent between 1913 and the end of the 1930s though the advance was unevenly distributed over time. After a spurt in the postwar boom, income per head fell below the prewar level in the ensuing depression, which was accentuated by a loss of income from abroad. From the middle of the 1920s a sustained rise set in with only a small dip during the depression years, and the overall gain in the 1930s (1930-32 to 1936-38) was of the order of 16.5 per cent. Most groups in society benefited from the rise in living standards, including the working classes, where trends in real earnings followed a similar pattern. Average annual real earnings rose by some 30 per cent over the period as a whole with much of the increase occurring during the war and postwar boom and in the early 1930s. These data make no allowance for non-monetary benefits such as the shorter working day, holidays with pay entitlement and better amenities and working conditions.

From the point of view of the matter in hand, the important issue is the trends in income and consumption through depression and beyond. Unlike the experience in some countries, for example the United States and Germany, real incomes and consumption in Britain did not collapse during the depression. In fact there was only a modest dip in total real national income and consumption during the depression years, after which both rose strongly. On the other hand, real wage rates and real wage earnings showed a substantial increase between 1928 and 1933, of the order of 13 per cent. Thereafter both levelled off and in the later 1930s remained fairly stable.

The remarkable increase in the real income of the working classes through depression can be attributed to two main factors: the stickiness of nominal wages and the shift in the terms of trade. Unlike the situation in the nineteenth century and even in the postwar contraction of 1920-22, wage rates were not battered down under the old sliding-scale principle. In fact nominal wage rates showed great resilience in the face of adversity; between 1928-33 nominal wage rates declined by just over 5 per cent while nominal wage earnings fell by slightly less. This increasing inflexibility of wages, which incidentally helped to exacerbate the unemployment problem (see below), can be attributed to several factors: the rapid expansion of unemployment insurance benefits and their coverage which set a floor to earnings, the regulation of certain wages by Trades Boards, and the growing strength and influence of the unions who were prepared to trade off unemployment against wages. Finally, the use of sliding-scale agreements to determine wages fell into disfavour after 1922.[26]

The second and more important determinant of the trends in real wage earnings was the marked improvement in the terms of trade. These had deteriorated after the 1921 recession until the mid-1920s when they stabilised, this being one reason for the sluggish real income growth in the first half of that decade. From 1928 they began to improve significantly as import prices plunged downwards. Between 1928 and 1933 import prices fell by 46 per cent, while export prices declined by less than one third, resulting in a terms of trade improvement of over one quarter. Britain stood to gain more than most countries from the dramatic collapse in primary product prices during this period because of her greater propensity to import foodstuffs and raw materials. The improvement led to a fall of nearly 16 per cent in the retail price index over the period as against a nominal wage rate reduction of around 5 per cent. Thereafter retail prices tended to edge upwards as the terms of trade improvement evaporated in the later 1930s following the strong recovery in primary product prices.

There were, in addition, longer-term forces at work making for an improvement in the real incomes of wage earners. One was the steady increase in productivity which over the long term may be seen as the main determinant of real income trends.[27] Secondly, there was an increase in the number of salaried workers, a narrowing of skill differentials and a constant wage-income ratio despite a fall in the proportion of wage earners to the total occupied population. Finally, there was a trend towards greater income equality through taxation. Whereas before the war the burden of taxation had generally been regressive (because of the high incidence of direct taxation) so that the working classes paid out more in tax than they received in benefits, this pattern was reversed in the interwar years as a result of the sharp increase in direct taxation during the war and the big extension of welfare benefits in the interwar years. The overall effect was that by the 1930s the post-tax incomes of the lowest income groups had gained by between 8 to 14 per cent, depending on the assumptions involved.[28]

The strong rise in real wage earnings throughout the depression was the main factor both in maintaining a high floor to consumption during the crisis period and for the sustained rise in consumption thereafter. Consumption was also bolstered by a rise in the propensity to consume, partly as a result of the shift in income distribution towards the lower income groups, and by the extensive system of welfare benefits, the coverage and real value of which rose considerably during the period despite the occasional economy cuts. *Per capita* expenditure in real terms on all social services increased more than threefold between 1913 and 1938,[29] while unemployment cash benefits in real terms advanced strongly. As against November 1920, when the vast extension of the unemployment insurance scheme took place, the real benefit received by a man with a wife and two children had risen by 240 per cent in 1931, and 92 per cent for a single man.[30]

The consumer-based recovery of the 1930s is therefore logically consistent with the trends in real income. The sustained rise in the real incomes of the wage-earning population helped to maintain consumption levels through the depression and subsequently it was translated into a rise in consumption which made the recovery effort effective. It is true that much of the income growth in the early 1930s was fortuitous, being derived from the terms of trade improvement and the real wage resistance on the part of the unions. However, be that as it may, the impact was nonetheless favourable and the evidence of shifting consumption patterns is plain to see. Since the majority in work were beneficiaries, their gains were translated into effective demand for houses, consumer durables and a wide range of services. It is clear that many working-class households were able to raise their discretionary spending during this period. There is certainly evidence that they were moving into home ownership, that they were buying more consumer products than ever before, and also spending more on entertainment and leisure. By the end of the 1930s the majority of households owned an electric iron and a radio or gramophone, one in seven had a vacuum cleaner and electric cooker and one in fifty had a refrigerator and washing machine.[31] The pattern of spending of the average working-class family also changed quite markedly during the period. Before 1914 some 60 per cent of the average household income went on food and another 16 per cent on rent and rates, leaving little for other necessities let alone the luxuries of modern living. By the end of the 1930s these two items absorbed only 44 per cent, while over the same period there was a sharp fall in the share of income spent on alcoholic beverages. This shift in spending patterns therefore allowed more to be spent on non-essentials, including fixtures and fittings in the home, furniture, consumer goods, newspapers, transport and leisure and entertainment, all of which showed significant gains in the interwar years.[32] The domestic economy was also aided by a switch in spending from imports to home goods in the early 1930s, the ratio of imports of goods and services to GNP (in current values) falling from 27 to 18 per cent between 1929 and 1932. Had the former ratio obtained in 1932, import values would have been £350 million higher, that is equivalent to some 7½ per cent of 1929 GNP.[33]

There is of course a danger of exaggerating the real income effects. Real income gains were small in the later 1930s and consumption through the cycle was no more buoyant than it had been in the later 1920s. But both can be considered crucial to the early years of recovery. The strong real income growth in the early 1930s and the high floor to consumption, coupled with the switch in patterns of spending, provided a strong platform for the initial lift off from the trough.

## Real Wages and the Profits Cycle

We now turn to the third and final natural force factor, which is a rather more complex issue. It relates to the reversal of the profitability cycle and the role of the real wage level in determining the cost-price structure of British industry, and its relevance to recovery. Longer-term there is also the question of the real wage level and unemployment to be considered. In view of the recent revival in neo-classical economic thinking and its re-application to the interwar period, it seems pertinent to examine the matter afresh.

The most positive stance on the role of real wages in the recovery has been put forward by Beenstock, Capie and Griffiths in their recent paper to a Bank of England symposium on recovery. They maintain that 'real wage behaviour after 1932 was highly conducive to job creation because the underlying growth rate in productivity was positive. Indeed, we shall argue that moderate real wage behaviour after 1932 played a major role in the economic recovery from the supply side.'[34] The crux of the issue is the movement of industry's cost-price structure as a determinant of profitability; this tends to get distorted at times of high economic turbulence when violent price changes occur, as in 1919-21, 1929-32, 1973-75 and 1979-81, with adverse consequences for profitability. What tends to happen is that industry's final output prices fall faster or rise more slowly than its input costs, the major one of which is wages. This adverse shift in the cost-price structure, reflected in a rise in own product real wages (wage costs relative to final output prices), leads to a collapse in profit margins which in turn reacts unfavourably on investment and employment. Recovery will not therefore take hold until the adverse trend in profits is reversed and this depends upon improvement in the cost-price structure either via real wage moderation or a firming up of product prices or both.

The problem was most acute in the tradeable goods sector subject to the world price level. Thus it was manufacturing which bore the brunt of the recession and where the job losses were heaviest. It seems likely that manufacturing industry was already under some strain before the depression set in since the relative price of manufactured goods had been declining during the 1920s, while after 1923 money wages remained remarkably stable and real wages actually rose. Hence own-product real wages deteriorated and profit margins came under pressure; the squeeze was partly offset by the rising volume of output and productivity growth in the later 1920s though these were insufficient to prevent a steady fall in the share of profits in GDP. Thus at the ruling level of prices, productivity and exchange rates, it could be argued that the real wage level was too high in the 1920s. But the position became very much worse with the onset of depression. Though raw material costs fell, nominal wage costs remained remarkably resilient at a time when manufacturers' output prices fell sharply, reflecting the depressed world price level for tradeable products. Thus wage costs in terms of output prices rose (a rise in the own-product real wage) and profit margins were torn to shreds. According to Beenstock *et al.*, the impact was particularly severe in 1931 when own-product real wages in manufacturing grew by 7½ per cent, whereas for the rest of the economy there was only a very modest increase. The reaction of manufacturers to this situation was as might be expected: stocks were rapidly reduced, output fell and the demand for labour and the level of investment declined. At the same time, higher real wages induced an increase in the labour supply through a rise in the participation ratio, which thereby exacerbated the unemployment problem from the supply side at a time when labour supply growth was quite buoyant due to the age structure of the population. The unwinding of the process came after 1932 when real wages began to moderate under the influence of labour glut, productivity and output rose and manufacturers' output prices began to firm. This allowed industry's cost-price structure to improve (own-product real wage fell), profits began to recover and the outlook became more favourable for investment and employment. At the same time the labour supply schedule began to flatten out as real wage growth moderated. Thus the British economy was pulled out of recession by the automatic reversal of the profits cycle as a result of the working of the labour market.

The logic of the analysis is attractive and it derives some empirical support from other periods in which distortions to the cost-price structure occurred. Industry's cost-price structure was badly impaired in the boom and slump of 1919-21, and similarly in the 1970s and early 1980s, with the result that profits collapsed and industry, exposed to the world price level, bore the brunt of the crisis. Only when profitability began to take a turn for the better through an improvement in its cost-price structure and a moderation in the real wage gap did recovery begin to take place.[35] On the other hand, the data on which the case for the early 1930s is based are far from impeccable. There is no satisfactory index for manufacturers' final output prices in this period and hence proxy prices have to be employed. In an earlier paper Beenstock used the retail price index, which is clearly unsuitable for the purpose because of its import content.[36] However, whichever deflator one uses to derive the own-product real wage (GDP deflator, retail price index or wholesale prices) the results emerge the same: that the real wage level rose strongly in depression and continued to do so until 1933-34. This would seem to be contrary to the proponents of the real wage hypothesis, unless it can be shown that the slowing down in real wage growth in 1932 and a simultaneous firming up of industry's product prices coincided with the recovery in profits. Unfortunately, the data are not sufficiently robust to substantiate this one way or another, though the shift of resources into non-traded goods, such as housing and services, in the early 1930s may reflect the more favourable cost-price structure in these sectors. The alternative explanation is that volume gains, productivity improvements and major cost-cutting exercises through labour shedding were initially more important in arresting the profits fall, after which real wage moderation helped to carry the economy forward.

Dimsdale's recent analysis of the relevant data reinforces doubts about the strength of the real wage argument in depression and recovery. While he confirms that when earnings in manufacturing are deflated by wholesale prices the behaviour of own-product real wages is consistent with the arguments advanced by Beenstock *et al.*, the use of more appropriate price deflators to eliminate some of the import content in the wholesale prices index produces rather unfavourable results. The rise in manufacturing-product real wages in the depression and their subsequent decline in recovery are much reduced, while the timing of the reversal does not occur until around 1934-35, which scarcely squares with the original argument. Even the proxy index used by Beenstock *et al.*, that of the price of exports and capital goods, does not produce a peaking of the own-product real wage until 1934, by which time recovery is well under way. A further point to consider is that Beenstock and his colleagues do not take account of the consequences for output of the fall in non-labour costs in this period. Raw material prices fell relative to those of manufactures in the depression and this was sufficient to offset the rise in labour costs so that aggregate input prices did not rise relative to output prices for manufactures. Conversely, in the recovery phase, 1932-37, moderation on the wage front was counter-balanced by a strong rise in the price of raw materials so that the weighted index of input prices increased by only slightly less than that of the industry's final output prices. 'On this evidence', Dimsdale concludes, 'it is not convincing to explain output variation in depression and recovery by reference to wage behaviour alone without taking account of the offsetting movements in prices of raw materials. Beenstock *et al.* have greatly over-emphasised the relationship between product wages and employment in the 1930s, in both slump and recovery.'[37]

If demand variables are included in the estimating equations the price effects become more significant. World trade and money had strongly expansionary effects and the elasticity of demand for labour with respect to wages for the whole of the interwar period is shown to be quite high. However, Dimsdale maintains that the contribution of the product wage to recovery was not large and that it assisted recovery only insofar as it was less contractionary than had been the case in the 1920s. 'Both the fall in employment, 1929-32, and the recovery, 1932 to 1937, are largely accounted for by the demand variables.'[38]

If the real wage hypothesis of depression and recovery is in some doubt, there is still a case for examining the real wage question for the period as a whole, since Dimsdale's own findings suggest that real wage effects on employment should not be minimised. This is particularly true for the 1920s as a result of the severe cost-price distortions which occurred in the war and immediately thereafter. By 1924 hourly real wages were 28 per cent higher than in 1913, while own-product real wages were some 20 per cent higher, as against a rise in output per head of only 5 per cent over the same period. Moreover, throughout the decade and into the early 1930s, manufacturers' output prices were falling faster than the combined index of labour and material input prices; between 1920 and 1929, for example, final product prices fell by nearly 50 per cent as against 39 per cent for input prices.[39] The questions one should ask therefore are: (1) was the real wage level too high? (2) did this exacerbate the unemployment problem? and (3) why did the real wage level not fall to clear the market?

It has frequently been alleged that the real wage level was too high in the interwar period given the prevailing cost structure of industry, the level of productivity and the ruling exchange rate, at least through to 1931. Keynes himself recognised that both the economy and the labour market had become less flexible than hitherto and that wage rigidity was a source of unemployment warranting government action. In 1930 he still believed that money wage reductions provided a solution to Britain's problems though his preference was for raising prices on the grounds of equity and feasibility. Hence his initial preference for a tariff which, by raising prices relative to wages, would stimulate employment.[40] Later writers have re-invoked the classical argument that the real wage level was too high. Dimsdale, for example, states that competitiveness could have been improved in the 1920s had the growth of real wages been moderated,[41] while the tenor of Casson's and Beenstock's recent work is that the demand for labour was dependent on the real wage level.[42]

The supposition that the real wage level is too high rests on the narrow premise of a *ceteris paribus* situation where all alternative variables are represented as an 'immovable feast'. This assumption is probably more or less valid for the 1920s given the decision on the exchange rate, the difficulty — in the short term at least — of influencing the level of productivity and the pattern of external demand. Real wages thus became the *deus ex machina* which determined the level of employment, a proposition which draws increasing strength from recent work in this field.[43] It follows therefore, that in conditions of underutilised resources a reduction of unemployment can be achieved by lowering the real wage level, which will stimulate the demand for labour and reduce its supply as entrants drop out of the labour force at the new ruling wage level.

The theory has a neat logic but in practice it is difficult to visualise, as with the exchange rate in the 1920s, a feasible equilibrium wage level which would have cleared the market. Casson's estimates for interwar manufacturing support a pre-Keynesian employment function, but they indicate that a sizeable downward shift would have been required to have had much impact; a one per cent fall in the own product wage would have increased employment by just over one third of one per cent, implying that a 10 per cent employment stimulation would have required a reduction of no less than 30 per cent in the own-product wage.[44] The size of the reduction would probably have been even greater in the depressed areas dependent on the staple industries. This of course does not take into account the effect on the labour supply nor, on the other hand, the demand effect through lower real wages. However, the orders of magnitude involved are too horrific to contemplate in political and social terms given the closeness of many wages to the subsistence level at this time.[45]

If the self-correcting mechanism working through real wages was not a very practical proposition, it still remains to be explained why the wage level was so unresponsive to the conditions prevailing in the interwar period. After the sharp decline in the early 1920s, nominal wage rates and earnings remained remarkably resilient to mounting employment. Even in the depression of the early 1930s the decline in nominal rates was very modest — around 5 per cent — while for the period as a whole there was a small rise, which given the intervening price fall meant a significant increase in the real wage level and in the own-product wage. Even more significant was the fact that industrial and regional wage differentials showed little tendency to increase in line with variations in unemployment rates.[46] How then does one explain the loss of wage flexibility in the wage structure?

Two factors in particular may be held responsible for the rigidity of nominal wages after 1922. The trade unions, despite a weakening in their strength after the General Strike, strongly resisted any downward pressure on the real wage level and for the most part they were successful. This approach entailed the implicit acceptance of high unemployment as the price to be paid for maintaining the wage levels of their members with jobs. At the same time they fiercely resisted any proposals to lower the benefits paid to the unemployed; in fact they were more anxious to secure an improvement in the conditions and rates of unemployment benefit than they were to find ways of increasing the opportunities for employment, and for obvious reasons. It relieved the unions of their traditional prewar role of responsibility for the maintenance of the unemployed, which now became the responsibility of the state. This meant that they no longer had to consider the unemployment implications of their wage policies as had been the case before the war, when a growing burden of unemployment in declining sectors meant acquiescence in wage cuts to relieve pressure on union funds paid out in benefit. Furthermore, a high level of welfare benefits eased the strain on the labour market in terms of job competition and left the unions free to concentrate on wage bargaining for the benefit of members in work.[47]

In effect therefore, the unemployment insurance system provided a safety net for those with jobs. The rising real value of benefits and their relatively high level — averaging 47 per cent of the weekly wage for the interwar years as a whole with a peak of nearly 60 per cent in 1938 — set a floor to the whole wage structure below which wages could not easily be depressed. Whether the unemployment insurance system was as powerful as Benjamin and Kochin have suggested in pulling workers into unemployment rather than their being pushed out of employment, is very much to be doubted.[48] The authors' contention that much of the unemployment in interwar Britain can be explained by the high benefit-wage ratio has been subject to a storm of criticism since it was first promulgated. Indeed, Casson has shown that an increase in unemployment benefits tended to reduce rather than increase the labour supply.[49] On the other hand, in terms of the demand for labour there may be more substance in the argument insofar as the level of benefits helped to prevent a downward adjustment of the wage level thereby depressing the demand for labour. Noreen Branson writing on the 1920s has commented thus: 'The existence of unemployment benefit, pushed up to something approaching the levels needed for maintenance and extended over many months for the long-term unemployed after 1924, greatly hindered the drive to bring down wages.'[50] The evidence on this point is not unequivocal but the example of the Poplar Board of Guardians, who paid relief above the ruling wage rates, is not by any means an isolated case. The Pilgrim Trust in their study of long-term unemployment estimated that possibly 20 per cent of those supported by unemployment assistance were at least as well off, if not better off, than had they been working,[51] while a survey of South Wales in the late 1930s showed that a third of single men and one half of married men were receiving more in unemployment allowances than they had in their last jobs.[52] Moreover, as Dimsdale notes, quoting Hicks' *The Theory of Wages* (1932), unemployment insurance may have kept up wages by dint of the fact that a worker could not be disqualified from receiving benefits by refusing work at wages below the standard rates negotiated by the unions.[53] The benefit system also discouraged mobility of labour by reinforcing the inherent inertia of workers to remain attached to existing communities and trades.

Whatever the impediments to greater flexibility in the wage

structure may have been in this period matters not too much, since in practical terms one must recognise the limitations to wage adjustment. No doubt wage adjustment could have cleared the market eventually, but at what level one dreads to contemplate. It may be that more than a few workers were receiving benefit at or above their last paid work, but this could simply illustrate how close to subsistence many wage levels actually were. Any significant compression of them would therefore have pushed many families into severe poverty or even starvation. As Glynn and Booth put the case: 'In the absence of unemployment benefit, it is possible that wage flexibility, greater regional wage variations, and higher labour mobility might have developed, but with potentially awesome consequences in social and political terms.'[54] In any case, the mismatch problems in the labour market were too large and too dispersed to be readily amenable to unilateral wage cuts; it is difficult to believe, for example, that wage reductions in the staple industries would have done much to staunch the flow of labour from these sectors, while the demand effects of wage cuts might have checked the growth of new developments in the areas in which they were located.

On the other hand, if the real wage level was too high and its rigidity maintained by real wage bargaining, then one has to recognise that not only did this make it more difficult to clear the labour market, but also that it set a constraint to policy action with respect to unemployment. Casson has demonstrated how the existence of real wage bargaining can result in pure 'crowding out' between the public and private sectors irrespective of the level of unemployment. Real wage resistance, when complete, ensures that the supply of output is inelastic with respect to prices since it is determined by the stipulated real wage, which adjusts in response to a rise in prices brought about by increased public demand, while at the same time real private demand will contract. Accordingly 'Once the real wage level has been fixed, expansionist fiscal and monetary policies will be purely inflationary: "crowding out" in the product market will be complete, and the employment multiplier will be zero.'[55]

In other words, the British economy in the interwar years was effectively locked into a low-level structural equilibrium trap from which there was no easy escape route. The position is not too dissimilar from that of Britain and Europe in the 1980s for which many solutions have been proposed, all of which have certain drawbacks. Whatever solution had been implemented in the interwar years, the process of adjustment would have been a long and painful one given the extent of the structural problems. Ironically, the solution, when it came, was more painful than any anticipated in peacetime.

## Recovery — A Final Word

The debate on the causes of recovery in the 1930s has been a long and arduous one partly because the matter has recently aroused so much interest due to the seeming parallels with today's conditions. Probably, as Dimsdale concludes, it would be a mistake to emphasise one factor to the exclusion of others.[56] It is clear that recovery was assisted by a combination of favourable factors, both on the policy side and in terms of spontaneous forces, though if pressed our own preference would still lie with the latter. And in particular we would emphasise two major factors: the strong thrust imparted by building and the new industries, and the demand effects of rising real incomes in the early years. These more than anything else were responsible for lifting Britain out of recession and sustaining the recovery until the later 1930s when rearmament became a force of no small importance.

## Notes

1. G. D. N. Worswick, 'The Sources of Recovery in U.K. in the 1930s', *National Institute Economic Review*, 110 (November 1984), p. 93.
2. G. D. N. Worswick, 'The Recovery of Britain in the 1930s', *Bank of England Panel Paper,* 23 (1984), p. 13 and 'The Sources of Recovery in U.K. in the 1930s', *National Institute Economic Review,* 110 (November 1984), p. 86.
3. N. von Tunzelmann, 'Britain 1900-45: A Survey', in R. Floud and D. N. McCloskey (eds.), *The Economic History of Britain since 1700.* Vol. 2, *1860 to the 1970s* (1981), pp. 249-51.
4. M. E. F. Thomas, 'Regional Unemployment and Policy in the 1930s: A Preliminary Study', Department of Economics, University of Essex (1981), p. 24.
5. T. Barna, 'The Interdependence of the British Economy', *Journal of the Royal Statistical Society,* A115 (1952).
6. The relationship between building and transport is explored fully in H. W. Richardson and D. H. Aldcroft, *Building in the British Economy between the Wars* (1968), Chapters 13 and 14.
7. Worswick appears to lay rather undue emphasis on the role of cheap money in the housing boom without considering fully the timing factor or the influence of other factors. See *National Institute Economic Review,* 110 (November 1984), pp. 90-1.
8. These factors are discussed at length in Richardson and Aldcroft, *op.cit.*
9. I. Svennilson, *Growth and Stagnation in the European Economy* (1954), p. 52

and W.W. Rostow, *The World Economy: History and Prospect* (1978), pp.228-9.

10. H.W. Richardson, 'The Basis of Economic Recovery in the Nineteen-Thirties: A Review and a New Interpretation', *Economic History Review*, 15 (1962-63), and *Economic Recovery in Britain 1932-39* (1967).
11. J.A. Dowie, 'Growth in the Inter-war Period: Some More Arithmetic', *Economic History Review*, 21 (1968). Dowie does conclude however that the new industries were probably the leaders in the upturn of the 1930s, his concern being through cyclical performance (1929-37).
12. S. Pollard, *The Development of the British Economy, 1914-1980*, (1983), p.54.
13. This would coincide with the Beenstock, Capie and Griffiths view, but conflict with Worswick's. See *Bank of England Panel Paper*, 23, (1984) and *National Institute Economic Review*, 110 (1984).
14. S. Glyn and A. Booth, 'Unemployment in Interwar Britain: A Case for Relearning the Lessons of the 1930s', *Economic History Review*, 36 (1983), pp.336-7.
15. Richardson later modified his conclusions on this point. H.W. Richardson, 'The Economic Significance of the Depression in Britain', *Journal of Contemporary History*, 4 (1969), p.12.
16. N.K. Buxton, 'The Role of the New Industries in Britain during the 1930s: A Reinterpretation', *Business History Review*, 49 (1979), pp.214, 215, 222.
17. The staple industries comprise agriculture and fishing, mining and quarrying, iron and steel, shipbuilding, mechanical engineering, cotton and woollen textiles; new industries include vehicle manufacturing, electrical engineering, rayon, non-ferrous metals, paper, printing and publishing. For an alternative definition see Pollard, *op.cit.*, p.55.
18. Buxton, *loc.cit.*, p.214.
19. S.N. Broadberry, 'Unemployment in Interwar Britain; A Disequilibrium Approach', *Oxford Economic Papers*, 35 (1983), pp.464, 468, 483.
20. R.C.O. Matthews, C.H. Feinstein and J.C. Odling-Smee, *British Economic Growth, 1856-1973* (1982), pp.256-8.
21. Glynn and Booth, *loc.cit.*, p.335.
22. G.N. von Tunzelmann, 'Structural Change and Leading Sectors in British Manufacturing, 1907-68', in C.P. Kindleberger and G. di Tella (eds.), *Economics in the Long View: Essays in Honour of W.W. Rostow:* Vol. 3, *Applications and Cases*, Part II (1982), pp.20, 46-7.
23. The industries included in the two sectors comprise building and contracting and building materials, vehicle manufacturing, electrical engineering, chemicals, electricity, gas and water, rayon, silk and hosiery. The respective shares for the new industries alone are 10.9, 14.9 and 18.2 per cent.
24. D.H. Aldcroft, 'Economic Growth in Britain in the Interwar Years: A Reassessment', *Economic History Review*, 20 (1967), p.321.
25. M.W. Kirby, *The Decline of British Economic Power since 1870* (1981), p.76.
26. See B. Henneberry, R. Keleher and J.G. Witte, '1919-1939 Reassessed: Unemployment and Nominal Wage Rigidity in the U.K.', Working Paper Series, Indiana University (no date).
27. E.H. Phelps Brown and M. Browne, *A Century of Pay* (1968), pp.167, 263.
28. U.K. Hicks, *Public Finance* (1947), p.299 and T. Barna, *The Redistribution of Incomes through Public Finance in 1937* (1945), p.214.
29. A.T. Peacock and J. Wiseman, *The Growth of Public Expenditure in the United Kingdom* (2nd edn., 1967), Tables A-15 to A-17, pp.184-91.
30. A. Deacon, *In Search of the Scrounger: The Administration of Unemployment Insurance in Britain, 1920-1931* (1976), p.17.

31. J.H. Davies, 'Intermission — the British Film Industry', *National Provincial* Bank Review, 43 (August 1958).
32. J. Stevenson, *British Society 1914-45* (1984), p. 126.
33. A. Ford, 'Unemployment — Lessons of the Inter-war Years', in Department of Economics (University of Warwick), *Out of Work: Perspectives of Mass Unemployment* (1984), p. 11.
34. M. Beenstock, F. Capie and B. Griffiths, 'Economic Recovery in the United Kingdom in the 1930s', *Bank of England Panel Paper*, 23 (1984), p. 65.
35. See D. H. Aldcroft, *Full Employment - the Elusive Goal* (1984). The real wage adjustment problem has been seen as particularly acute in Europe during the last decade. M. Emerson (ed.), *Europe's Stagflation* (1984).
36. M. Beenstock, 'Do U.K. Labour Markets Work?', *Economic Outlook* (LSE) 25 (Jan./July 1979).
37. N.H. Dimsdale, 'Employment and Real Wages in the Inter-war Period', *National Institute Economic Review*, 110 (November 1984), p. 98.
38. *Ibid.*, p. 102.
39. *Ibid.*, pp. 94, 98.
40. B.J. Eichengreen, *Sterling and the Tariff, 1929-32* (1981), pp. 7-12.
41. N. H. Dimsdale, 'British Monetary Policy and the Exchange Rate, 1920-1938', *Oxford Economic Papers, Supplement* (1981), p. 345.
42. M. Casson, *Economics of Unemployment* (1983), p. 189 and Beenstock, *loc.cit.*
43. See Casson, *op.cit.*; D. Grubb, R. Layard and J. Symons, 'Wages, Unemployment and Incomes Policies', in M. Emerson (ed.), *Europe's Stagflation* (1984); Oxford Economic Forecasting Group Enquiry, reported in *Financial Times*, 1 November 1984, p. 7. More generally, R. Layard and J. Symons, 'Do Higher Real Wages Reduce Employment', *The Economic Review*, 1 (January 1984), pp. 16-19.
44. Casson, *op.cit.*, p. 206.
45. B. Sadler, 'Unemployment and Unemployment Benefits in Twentieth Century Britain: A Lesson of the Thirties', in Department of Economics (University of Warwick), *Out of Work: Perspectives of Mass Unemployment* (1984), p. 28.
46. S. Glynn and A. Booth, 'Unemployment in Interwar Britain: A Case for Relearning the Lessons of the 1930s', *Economic History Review*, 36 (1983), p. 347.
47. J. A. Garraty, 'Unemployment during the Great Depression', *Labour History*, 17 (1976), pp. 137-8.
48. D.K. Benjamin and L.A. Kochin, 'Searching for an Explanation of Unemployment in Interwar Britain', *Journal of Political Economy*, 87 (1979).
49. Casson, *op.cit.*, pp. 208-9; for some of the criticisms of Benjamin and Kochin's thesis see the series of articles in the *Journal of Political Economy*, 90 (1982); T. Hatton, 'Unemployment Benefits and the Macroeconomics of the Interwar Labour Market: A Further Analysis', *Oxford Economic Papers*, 35 (1983); Sadler, *loc.cit.*
50. N. Branson, *Britain in the Nineteen Twenties* (1975), p. 249; cf. Casson, *op. cit.*, pp. 208-9.
51. Pilgrim Trust, *Men Without Work* (1938), p. 120.
52. J. Stevenson, *British Society, 1914-45* (1984), p. 282.
53. N. H. Dimsdale, 'British Monetary Policy and the Exchange Rate, 1920-1938', *Oxford Economic Paper, Supplement* (1981), pp. 242-3.
54. Glynn and Booth, *loc.cit.*, p. 347.
55. Casson, *op.cit.*, pp. 79, 83, 249.
56. N.H. Dimsdale, 'Employment and Real Wages in the Interwar Period', *National Institute Economic Review*, 110 (November 1984), p. 102.

# 7 The War Economy and its Consequences

The decade of the 1940s was undoubtedly a period of severe austerity for the British people. Shortages and rationing were the dominant themes and rich and poor alike shared in the privation. Unlike the 1930s however, it was a time of full employment and of greater equality in income and consumption, which no doubt helped to ease the experience of common sacrifice. This chapter considers some of the main aspects of the war economy and its consequences, while the final chapter deals with the issues of reconstruction and recovery under the Labour Governments of 1945-51.

It would be tedious to describe in detail the elaborate machinery of wartime planning, control and resource allocation as it developed through the various stages of the war, all of which have been extensively covered in the excellent multi-volume civil history of the period.[1] This chapter therefore adopts a selective approach, looking first at the overall extent of mobilisation of the economy and the achievements of the productive machine, and then moving on to the means by which it was made possible through the control apparatus and the arrangements to finance it. Finally, the chief economic consequences in terms of the reconstruction tasks ahead are discussed.

## Extent of Economic Mobilisation

At the outbreak of war in September 1939 Britain was somewhat better prepared administratively than she had been at the same stage in 1914. Several government departments and official committees had been deliberating and making preparations since the early 1930s for the likely event of another war, drawing on the experience of the previous period of hostilities. From their work it became apparent that a more extensive and effective system of economic mobilisation and control would be required in the event of total war than that obtaining in the first world war. This conclusion was reinforced in the later 1930s in the light of Germany's lead in rearmament and the strains which the British government's rather more modest effort was imposing on the economy. Indeed, it soon became evident that Britain would have to make haste in the matter of mobilisation if she were to avoid disastrous consequences. It was also acknowledged that Britain would not be able to wage a prolonged European war without assistance from her Allies.

On the question of the extent of Germany's mobilisation and the precise strategic aims of that country — Blitzkrieg or long war — there has been considerable debate both at the time and since.[2] Contemporary reports no doubt exaggerated the depth of German rearmament in the 1930s, but there seems little reason to doubt the claim that prior to the outbreak of hostilities Germany was on a stronger war footing than Britain. Throughout the years 1935-38 Germany had been devoting more than twice the proportion of her GNP to defence spending than had Britain, while at the same time she had assembled a fairly elaborate, if at times somewhat inefficient, system of controls over the economy to prevent the accelerated economic effort from short-circuiting the economy.[3] The striking contrast between the two countries in these respects, which goes some way to explaining the initial successes of Germany in the early stages of the war, reflects the differing policy approaches of the respective regimes in the 1930s. As Hancock and Gowing observed, 'the economy of the Reich, fully employed and firmly controlled, was already geared to war, while the economy of the United Kingdom was still far from full employment and only beginning to disentangle itself from the doctrine of "normal trade".'[4]

In view of this discrepancy the speed with which Britain got her act together is all the more remarkable. The more so in that the turning point did not occur until early 1939 when the Cabinet reversed its earlier policy priority of domestic stability over defence. Henceforth Britain's defence programme rolled forward at a prodigious pace so that in the financial year 1939 defence spending had almost caught up with that of Germany in relative terms — 21.4 per cent as against 23.0 percent of respective GNPs. By the start of the war Britain's arms production already matched that of her rival and in several important areas, for example aircraft and tank production, the German level of output had been surpassed. By the following year Britain's total defence effort exceeded the German by a considerable margin, accounting for over half the national income, and it was to remain around this level for the rest of the war.[5] Apart from the first year of hostilities defence expenditure absorbed over 80 per cent of total government outlay and Britain's effort was proportionally greater than that of any other combatant nation.

A second notable achievement was secured in the field of labour mobilisation. Two essential tasks were involved in this respect. The first was to ensure an adequate supply of labour overall, and the second involved reallocating labour to essential needs, that is away from civilian production to the armed forces and into munitions production. Even before the war, despite high unemployment, labour shortages had been evident in certain sectors, notably skilled workers in engineering, aircraft production and construction, and this problem became more acute once the war had begun. Subsequently an overall shortage of labour emerged once the unemployed had been mopped up. Thus from 1942 onwards comprehensive manpower budgeting became the main instrument in planning war production, and the allocation of labour among competing needs was a vital factor in maintaining the war machine. Reallocation of labour towards essential war needs including the combatant services, was effected through conscription and industrial direction, though the latter power was used as sparingly as possible in the case of civilian employment for fear of provoking industrial unrest as had been the case in the first world war. Recruitment to the armed forces was governed by the four National Service Acts passed between 1939-42, while a fairly extensive network of controls was instituted over civilian manpower under the Emergency Powers legislation of 1939. The latter gave the Minister of Labour and National Service the right to issue regulations to control the use of all labour, his powers including those of registration, direction and the regulation of engagement and dismissal of workers. By the middle of the war all adults up to the age of 50 had been registered and some 8 million workers were tied to their jobs by essential work orders.

The overall success of the manpower planning exercise can be judged by the fact that at the peak of labour mobilisation some 10 million persons, or about one half the total labour force, were either in active service or employed on munitions work, while even as late as June 1945 the proportion was still around 45 per cent. At that date only two per cent of the labour force were producing for export as against nearly 10 per cent prewar; eight per cent were providing and maintaining the nation's capital stock and the remainder were engaged in producing goods for civilian consumption.[6] The changes in the distribution of manpower between 1938-43 can be seen from Table 1. The main beneficiaries were the metal trades, engineering,

*Table 1: Distribution of Total Manpower in Great Britain (000s), 1938-1943*

| | June 1938 | June 1943 | Percentage Change |
|---|---|---|---|
| Allied Forces and Auxiliary Services | 385 | 4,761 | 1,136.6 |
| Total in civil employment | 17,378 | 17,444 | 0.4 |
| Metals, engineering, vehicles and shipbuilding | 2,590 | 4,659 | 79.9 |
| Chemicals, explosives, paints, oils | 276 | 574 | 108.0 |
| National and local government, including civil defence | 1,386 | 2,109 | 52.2 |
| Agriculture and fishing | 949 | 1,047 | 10.3 |
| Mining and quarrying | 849 | 818 | -3.7 |
| Textiles | 861 | 669 | -22.3 |
| Clothing, boots and shoes | 717 | 493 | -31.2 |
| Food, drink and tobacco | 640 | 519 | -18.9 |
| Cement, bricks, pottery, glass, etc. | 271 | 170 | -37.3 |
| Leather, wood, paper, etc. | 844 | 539 | -36.1 |
| Other manufactures | 164 | 123 | -25.0 |
| Building and civil engineering | 1,264 | 726 | -42.6 |
| Gas, water and electricity | 240 | 200 | -16.7 |
| Transport and shipping | 1,225 | 1,176 | -4.0 |
| Distributive trades | 2,882 | 2,009 | -30.3 |
| Commerce, banking, insurance and finance | 414 | 282 | -31.9 |
| Miscellaneous services | 1,806 | 1,331 | -26.3 |
| Registered insured unemployed | 1,710 | 60 | |
| Total working population | 19,473 | 22,285 (a) | 14.4 |

(a) Includes 20,000 ex-members of Forces in employment.

*Source:* Central Statistical Office, *Statistical Digest of the War* (1951), p. 8.

shipbuilding and chemicals and allied trades, where there was an overall increase in numbers employed of over 80 per cent, while those employed in government service increased significantly. In virtually all other industries there was a loss of manpower, in some cases of more than a third.

By 1942 an absolute shortage of labour for both civilian and war production had become a pressing problem. Virtually all the prewar unemployed had been absorbed into the active labour force (unemployment was down to an irreducible minimum of 0.5 per cent), while the armed forces were continuing to drain men away to the front. Four million were already engaged in the fighting services and before the end of the war approximately a million more were to swell their ranks. Thus the numbers producing at home dwindled steadily from the peak in 1939, and by 1945 the total in civil employment had fallen by over 2½ million.[7] The shortfall would have been even greater but for major additions to the active labour force through taking workers out of retirement, the mobilisation of women and the extension of working hours to compensate for the loss in numbers. At the peak the first category probably totalled some one million though many no doubt worked on a part-time basis. Much more important was the drafting into active industrial service of some 2½ million women, mainly housewives, domestic workers and those not previously employed, which did much to ease the overall labour scarcity. Britain went farther than any other country in utilising its womanpower in time of war and it was the only country which actually had the power to conscript or direct women into essential war production. The proportion of women aged over 14 employed in Britain rose from 27 to 37 per cent between 1939-43,[8] and by the latter date women were working in large numbers in occupations formerly deemed to be male preserves. The problem of skill shortages remained throughout the war, partly due to the difficulty of ensuring that key workers did not desert to the front. However, the skill bottlenecks were eased in various ways, for example, by accelerated training schemes, the careful control and allocation of skilled craftsmen and the dilution of work practices by redesigning jobs, with the aid of new machinery in some cases, so that unskilled men and women could take over tasks formerly reserved for skilled employees.

Another area which required fairly comprehensive mobilisation was the field of transport and trade. Because of the need to economise on the use of scarce resources such as food, raw materials and fuel, and to save foreign exchange, it was essential to establish tight control over foreign trade. Similarly in the case of transport, since the pressure on facilities for the movements of troops and essential supplies meant that transport space was at a premium. Accordingly, as in the first world war, the government took control of the railways at the start of the war and vested operational responsibility in the hands of the Railway Executive Committee (composed of the general managers of the four main companies), whose task was that of running the railways and London Transport as a unified whole with the obligation to give priority to all government traffic. Other forms of inland transport were subsequently controlled but it was the railways which bore the brunt of wartime transport requirements within Britain. Despite steps to eliminate all non-essential demands for both freight and passenger traffic, the total volume of traffic soared. By 1944 freight traffic was nearly 50 per cent greater than prewar while passenger traffic had almost doubled by the following year.[9] All this was accomplished despite a decline in rolling stock and a neglect of repair and maintenance work and renewal of facilities, principally through more efficient utilisation of the existing capacity as a result of unified operation.

The requisitioning of shipping space took somewhat longer but by the beginning of 1940 the newly established Ministry of Shipping was well on its way to taking control of all vessels on the UK and Colonial registers which were engaged in the deep sea lines and tramp trades. By the time France had fallen, only 10 per cent of the ships on the combined registers continued to trade free.[10] The government also became a significant shipowner in its own right and it had a number of ships on requisition or charter from other countries. Rates of remuneration were fixed by the government — somewhat lower than the generous scales allowed in the first war — and requisitioned vessels were effectively at the service of the state.

There were good reasons why it was deemed necessary to mobilise shipping space rapidly and comprehensively. The piecemeal and partial requisitioning of the previous war had proved to have serious limitations. Secondly, the prospects of greater damage to shipping through aerial warfare in the second world war also prompted firm action. Thirdly, the pressure on space was also likely to be greater because of the dispersed nature of the battle zones.

It was the need to save scarce shipping space together with that of conserving foreign exchange which forced the government to take comprehensive control of imports. In June 1940, a general control of imports was imposed and by the end of that year practically all imports into the UK were either brought in under licence issued by the Board of Trade or bought directly on government account through the various ministries responsible. By the end of the war little freedom was left to the private trader; less than 3 per cent of total imports were admitted completely unrestricted. The bulk of Britain's imports (nearly 63 per cent) was bought on government account, while the remainder was subject to strict import licensing.[11] As a result the volume of imports was cut back by nearly 40 per cent during the course of the war.

Mobilisation and control of resources were not of course confined solely to the areas discussed above. The exigencies of total war demanded a degree of mobilisation and control beyond anything ever previously experienced. The controls over resource use in the second world war were far more numerous, far deeper and probably more efficiently administered than those of the first world war. From the very beginning, under emergency legislation promulgated at the outbreak of war, the state armed itself with the widest powers and it subsequently used these to draw a tight network of controls around production, consumption, investment, distribution, transport, building work, foreign exchange, imports, manpower, prices, raw materials and many industrial goods.[12] In fact at the termination of hostilities there was hardly a thing which had not been mobilised or controlled in one way or another for the service of the war machine. As Youngson recalls, total war entailed 'that everything that could usefully be sacrificed in order to build up and maximize the military strength of the country for the time when it was possible to strike back against Germany in Europe (in 1944-45) was sacrificed.'[13] Total war in effect meant total mobilisation of the nation's resources, and in this respect Britain performed far better than her major adversary.

## The Productive Effort

Overall the economy performed very creditably throughout the war years, especially in view of the severe shortages of resources of all kinds and the enormous dislocation caused by aerial warfare. The aggregate output of the economy increased steadily through to 1943 after which it tailed off. At its peak it was some 27 per cent above the last year of peace. This could of course in no way match the massive leap in activity recorded in the United States where total output nearly doubled during the period, nor for that matter the 80 per cent increase in Canada (1938-44). But against most European countries it compared very favourably, especially in the later stages of the war when output fell back sharply in several Axis and occupied powers (Germany, Austria, Italy, France and Belgium). Britain did not therefore emerge from the war in a prostrate condition with a level of activity well below prewar, though naturally enough there were serious distortions to the former peacetime product mix, while exports had to be sacrificed to the war effort. At the nadir in 1944 the volume of exports was about 70 per cent down on 1938 which itself was a slack year (73.5 per cent down on 1937 and only 17.3 per cent of the 1913 level).

Some idea of the enormous increase in war and war-related products can be seen from the figures in Tables 2 and 3. The indices of munitions production for selected items (Table 3) clearly show that output of each product rose severalfold during the course of the war. In addition, there were some spectacular rises in the production of the heavier items of war for example aircraft, aircraft engines, tanks, bombs and artillery. Thus the number of aircraft produced rose more than ninefold, bombs by sixfold, tanks by more than twenty times and aircraft engines by a factor of ten. Many branches of engineering were given a new lease of life with the demands of war. The machine-tool industry is a good case in point. A somewhat backward sector prior to the war, it was transformed out of all recognition by the tooling required for war work. Production, which had languished around 20,000 machine tools in 1935, rose to 35,000 in 1939 and then jumped to 100,000 in 1942, while employment rose more than threefold during the same period.

The main beneficiaries in terms of expansion and modernisation were inevitably those industries and activities most directly related to the war effort.[14] Thus aircraft production, electrical and general engineering, metal working and chemicals and allied trades experienced the largest physical increases in activity, while agriculture,

*Table 2: Index of Ministry of Supply Munitions Production 1939-1944*

| | Average Sept.-Dec. 1939 | 1942 | 1943 | 1944 |
|---|---|---|---|---|
| Guns, small arms and instruments | 100 | 677 | 752 | 428 |
| Shells and bombs filled | 100 | 965 | 720 | 538 |
| Small arms and ammunition | 100 | 2623 | 4778 | 4607 |
| Propellants and high explosives | 100 | 529 | 503 | 574 |
| Armoured fighting vehicles | 100 | 1700 | 1671 | n.a. |
| Signal | 100 | 487 | 814 | 807 |
| Engineer and transportation | 100 | 582 | 1001 | 1635 |

*Source:* Central Statistical Office, *Statistical Digest of the War* (1951), p. 139.

*Table 3: Output of Selected Products in Great Britain 1938-1944*

| | 1938 | 1939 | 1940 | 1941 | 1942 | 1943 | 1944 |
|---|---|---|---|---|---|---|---|
| Aircraft (Number) | 2,828 | 7,940 | 15,049 | 20,094 | 23,672 | 26,263 | 26,461 |
| Aircraft bombs (short tons) | — | — | 51,093 | 147,848 | 211,048 | 223,807 | 309,366 |
| Tanks and self-propelled tanks (Number) | 419 | 969 | 1,399 | 4,841 | 8,611 | 7,476 | — |
| Aircraft Engines (Number) | 5,431 | 12,499 | 24,074 | 36,551 | 53,916 | 57,985 | 56,931 |
| Wireless Sets (Number) | 3,567 (a) | — | 19,616 | 26,015 | 101,145 | 193,076 | 144,161 |
| Coal (Mn. tons) | — | 231.3 | 224.3 | 206.3 | 204.9 | 198.9 | 182.8 |
| Steel (000 tons) | 10,398 | 13,221 | 12,975 | 12,312 | 12,942 | 13,031 | 12,142 |
| Iron Ore (000 tons) | 11,859 | 14,486 | 17,702 | 18,974 | 19,906 | 18,494 | 15,472 |
| Electricity (Mn. Kilowatt hours) | — | 27,733 | 29,976 | 33,577 | 36,903 | 38,217 | 39,649 |
| Cotton Yarn (Mn. lbs.) | 1,234 (b) | 1,092 | 1,191 | 821 | 733 | 712 | 665 |
| Tractors (Number) | 10,679 | 15,733 | 19,316 | 24,401 | 27,056 | 25,059 | 23,022 |
| Ploughs (Number) | 12,580 | 16,665 | 23,172 | 24,657 | 21,414 | 19,246 | 23,701 |

(a) Average April 1936 - August 1939
(b) 1937

*Source:* Central Statistical Office, *Statistical Digest of the War* (1951), pp. 75, 133-62.

another crucial sector, responded very satisfactorily. On the other hand, coal mining and iron and steel, both vital to the war effort, fell well below expectations for special reasons.

The vast increase in war production could not of course have been achieved without a very significant shift in the nation's product mix. Wherever possible resources were shifted out of civilian production and exports and into military production. The changing pattern of employment gives some indication of the magnitude of the shift (Table 1). Apart from the armed forces, which at their peak absorbed some five million persons, the two main areas of expansion were metals, engineering, vehicles and shipbuilding, and the chemicals, explosives and allied trades; their combined employment rose by over 80 per cent between 1938-43 to reach a total of over five million at the peak. The only other significant expansion occurred in government service, where numbers increased by more than one half to reach a total of over two million in 1943. In fact the war spawned a veritable army of non-industrial bureaucrats many of whom were co-opted on a part-time basis.[15] By contrast, nearly all other forms of activity including coal mining lost manpower, in some cases very significantly. In several sectors of the economy, for example, construction, building materials, leather, wood and paper, the distributive trades, commerce, banking and finance, the drop in employment amounted to 30 per cent or more between 1938 and the peak of mobilisation as manpower drained away to feed the armed forces or the industrial war machine.

The bulk of the employment and output contraction occurred in those industries producing mainly for the civilian market or for export. The worse-hit sector must surely have been housebuilding which virtually came to a standstill in the later years of war. In England and Wales less than 6,000 houses a year were being constructed in the final years of war as against an annual average of 334,405 between 1935-1938.[16] But the output of many civilian goods industries was also sharply reduced; cotton yarn production declined by nearly 53 per cent between 1938 and 1945, while the output of a wide range of consumer goods was severely curtailed. The result was that many household goods, including furniture, became very scarce and the same could be said of more essential commodities such as clothing and boots and shoes, while what was available tended to be of a standard variety and poor in quality. Inevitably the restriction of civilian production and rationing circumscribed the range of consumer choice, as a result of which spending declined during the war (see below).

Limitations on inessential production and rationing were not the only ways of securing a rise in the output of industries servicing the war machine. Long hours of work and the recruitment of a vast army of women into munitions work made a significant contribution to boosting production. In addition, there was some increase in efficiency, more particularly in those industries undergoing rapid expansion. For the whole economy output per worker rose by some 10 per cent between 1938-43 after which it declined, which is probably quite creditable given the fact that capital per worker fell continuously to a low point of 13 per cent less than the 1938 level.[17]

Two industries — agriculture and coal mining — are worth considering in some detail since both were of crucial significance to the war effort yet their response was vastly different. Agriculture put up a splendid performance and did much to save both shipping space and foreign exchange, while at the same time helping to ensure that the nation was reasonably well fed. The coal industry, on the other hand, continued to reflect its prewar depressed state and production and productivity declined throughout the war years.

Before the war agriculture had been a relatively depressed sector though the government's protective policy in the 1930s had secured some modest relief. At the time Britain depended heavily on foreign sources for its food supplies, but it soon became evident once war broke out that the degree of dependence would have to diminish, partly because of the physical problems involved in ensuring adequate supplies from abroad, but also because of the excessive demands it placed on shipping space and foreign exchange earnings both of which would be in short supply. Food, fodder and fertilizer imports, for example, accounted for some 23 million tons of shipping space in 1938. However, given an overall resource shortage at home, especially of manpower, the key thing in agriculture was to maximise output in terms of inputs while at the same time ensuring the production of those items which had the highest calorific value in terms of human consumption. Such considerations therefore dictated, as in the first world war, a switch from pasture to arable farming, thereby reversing a peacetime trend of many years standing, and in particular a concentration on filler products, wheat, potatoes and animal fodder, which gave high calorific values, were economical in the use of labour and at the same time saved both foreign exchange and shipping space.[18]

These objectives formed the basis for the transformation of British farming during the war. The shift in land use was impressive as the figures in Table 4 demonstrate. The arable acreage rose by nearly 50 per cent, from just under 12 million acres in 1936-38 to nearly 18 million by 1944, with a corresponding decline in permanent grassland, while the total tilled area rose by some 5 million acres. The main arable extensions occurred in wheat, barley, oats, corn, potatoes, vegetables and fodder crops (Table 4). Conversely there was a sharp fall in the number of animals: the number of sheep and lambs declined from 25,993 thousand in 1939 to 19,435 in 1944, pigs from 3,822 thousand to 1,571 (1938-43) and poultry from 64,053 thousand to 35,229 (1938-44). Only the number of cattle was maintained or slightly increased in an effort to improve the domestic milk supply. As a consequence, by 1943 meat output had fallen by over 30 per cent compared with 1938 whereas wheat production had increased by some 200 per cent.

Altogether net agricultural output probably rose by some 35 per cent at the peak of wartime production, though because of the changes in the product mix the increase in calorific value was more of the order of 70-90 per cent higher than prewar. There was also a notable increase in yields per acre in most of the main crops: in wheat from 17.7 cwts on average 1936-38 to a peak of 20.4 in 1942; in barley from 16.4 to 18.9 cwts over the same period; oats from 15.7 to 17.1 cwts, rye grain 12.5 to 15.4 cwts; and potatoes from 6.9 to 9.7 tons in 1940. Yields tended to decline generally in the later stages of the war.[19] There was also an increase in the productivity of labour as well as a modest increase in the size of the labour force, the loss of workers to the services being compensated by recruitment through the women's land army scheme.

The farming community did not accomplish this task unaided. In fact farmers did very well out of the war and agriculture became one of the most subsidised sectors of the economy as a result of the government's vigorous agricultural programme combined with a guaranteed market.[20] The main stimulus to the switch in production was the ploughing up subsidy introduced in 1939 coupled with the continuation of the prewar guaranteed price system for arable products, which became regularised in the form of annual price reviews. As a result agricultural prices rose much faster than input costs or prices in general so that in terms of net incomes farmers did very much better than other groups in society. In addition, they also received numerous subsidies for improvements and land reclamation, for mechanisation and the use of fertilizers as well as financial assistance for specific livestock such as cattle and sheep, and crops such as wheat and potatoes. This extensive programme of aid did produce some solid long-run benefits. There was a sharp rise in the use of fertilizers and a rapid spread in mechanisation. The diffusion of machines and implements during the war was probably greater than in the whole of the

*Table 4 Acreage devoted to Crops, 1936-38 to 1944 (000 acres)*

| | 1936-38 | 1944 |
|---|---|---|
| Total tilled area, of which | 8,471 | 13,708 |
| Wheat | 1,851 | 3,214 |
| Barley | 926 | 1,957 |
| Oats | 2,130 | 3,215 |
| Mixed corn | 96 | 415 |
| Potatoes | 597 | 1,219 |
| Sugar beet | 335 | 431 |
| Vegetables | 277 | 499 |
| Fodder crops | 1,435 | 1,971 |
| Total arable | 11,991 | 17,936 |
| Permanent grassland | 17,368 | 10,809 |
| Temporary grassland | 3,520 | 4,228 |
| Rough grazings | 15,944 | 16,278 |

*Source:* Central Statistical Office, *Statistical Digest of the War,* p.57.

interwar period. The number of tractors alone more than tripled between 1938-45, from 60,000 to 190,000, while there was a significant advance in the use of a whole range of machinery including disc harrows, cultivators, binders, havester-threshers, potato spinners and milking machines.[21] In fact British agriculture became the most highly mechanised and efficient in the world, with output per man year rising by as much as 10-15 per cent.[22]

Thus if the nation had to pay a price for securing its food supply in wartime, it was probably well worth paying. Increasing self-sufficiency meant enormous savings in foreign exchange and shipping space, while the nation was assured a basic diet, filling if not always very palatable.[23] By early 1943, for example, the Ministry of Food's import programme was running at an annual rate of only 6 million tons compared with an original projection of 22.5 million tons.[24] Altogether about one half the food import requirements was saved, with a switch from bulky cereals and cattle fodder to imports of meat and dairy products. However, the longer-term consequences of the wartime support programme were that agriculture secured a privileged position which it retained in peacetime in the form of guaranteed prices and deficiency payments.

By contrast, the coal industry was rather a disaster story during the same period. It was virtually the only industry of significance to the war effort which failed to come up to expectations. Coal output actually fell in every year of the war and by 1945 the industry was producing about one quarter less than 1939, at 174.7 million tons (excluding open-cast coal mining). Initially the decline in output was not a serious disaster, especially after the fall of France in 1940, since this relieved the industry of a major source of demand. However, by 1942 consumption was beginning to outstrip supply and had it not been for several offsetting factors a coal shortage would have become a serious impediment to the all-out war effort. The first of these was the development of open-cast coal which by 1944-45 was producing over 8½ million tons compared with negligible quantities at the start of the war. Secondly, the consumption of coal for certain uses was severely curtailed to allow for the increased needs of electricity works, the railways and the service departments. The household user of coal bore the brunt of the cuts, domestic house coal consumption being reduced by about one quarter between 1938-44, while industry suffered a 10 per cent cut from December 1943 when a programmed allocation system was introduced. The third factor which helped to relieve the situation was stock withdrawal. Stocks of coal were allowed to run down sharply after 1942-43 when the deficiency in current output was beginning to bite severely, and by the end of the war they had been reduced to about 10 million tons against a normal peacetime level of around 17 million. By then, with production at an all-time low and stocks depleted, the prospects for peacetime coal supplies were none too bright.

But why did the coal industry perform so badly? Several factors were responsible including the unhappy state of the industry prior to the war. The first problem was a loss of manpower to the services. In the early stages of the war this had not proved to be a major setback since the losses were more or less made good by new recruits. But with the fall of France a significant source of demand was removed which resulted in the temporary unemployment of some miners, who took the opportunity to join up following the raising of the restriction on the age reservation for military service from 18 to 30 in the autumn of 1940. The result was a sharp decline in the labour force and by the end of 1941 the coal industry had lost some 11 per cent of its 1938 workforce, including natural wastage. Subsequently employment in the industry stabilised as efforts were made to revive recruitment.

Manpower losses were not the only problem however. After 1939 productivity in the industry declined steadily. Annual output per wage earner fell from 301.9 tons in 1939 to 259.2 in 1944, a drop of 14 per cent. This was largely a product of two factors. First a decline in the proportion of miners working at the actual coal face in the early years of the war; secondly, after the middle of 1941 it was mainly the result of a serious fall in output per manshift amounting to between 10-12 per cent by 1944. This deterioration can be attributed to an ageing workforce on 5½ shifts or more per week, a run-down of plant and equipment despite some advances in mechanisation, and discontent over wages and working conditions including the use of the Essential Work Order to check absenteeism. It was part of an old story of poor industrial relations and a militant workforce. In wartime as in peace the coal industry retained its reputation for being one of the most discordant industries. The mines in fact accounted for one half to two thirds of the total working days lost in strikes in the years 1943 to 1944, which is estimated to have cost four million tons of coal in lost production. An added problem was the inefficient planning and organisation of mining activities due to the government's reluctance to institute strong central control over the industry.[25]

Apart from the major exception of coal, the wartime productive effort can be considered a remarkable feat.[26] The prodigious expansion in the output of war goods and certain types of food within such a short space of time clearly demonstrates the flexibility of the economic machine once the exigencies of the time necessitated a major reorganisation of the production structure and the reallocation of resource use. Even so, Britain's needs could not have been met unaided. American lend-lease played a vital part in supplementing supply in both armaments and food materials, as did the resources of the British Empire. Nor could Britain have devoted so large a share of her own resources to the war effort without a severe squeeze on domestic resource use through control and allocation. To these subjects we now turn.

## The Control of Resource Use and Consumption

It would have been impossible for Britain to have achieved such a remarkable effort on the economic front had the state not instituted centralised organisation and control of the nation's resources. Yet though the state soon became the largest single importer and the largest single buyer of the products of the war machine, it did not for the most part acquire ownership of the means of production, even though the powers of control assumed at the beginning of the war were almost limitless. Thus apart from the Royal Dockyards and the Royal Ordnance Factories which the state already owned, together with a number of agency factories acquired during the war, the state did not own or take over the essential means of production. Nor did it requisition or commandeer directly many activities apart from some basic services such as shipping, the railways and the ports. For the most part private business was left in control, though subject of course to a wide range of detailed controls on resource use, prices, investment and the like.

By the end of the war the range of controls was very wide and few products or activities were left completely free. Many of the controls had been applied in the first world war though on a rather more limited scale and with no overall financial plan or comprehensive programme of resource use.[27] Their re-application in the second war brought many technical improvements. For example, the gaps were plugged in the framework of price control by the implementation of a price stabilisation scheme and the injection of subsidies. An attempt was also made to calculate the inflationary gap. Excess profiteering was avoided by more rigorous taxation and tight cost control. Furthermore, a number of controls introduced in this period had never been used in the 1914-18 war, notably the concentration of civilian industries, utility schemes, points rationing, the direction of labour, forced loans from the public and exchange control. In the last case it is true that a rudimentary form of control over the foreign exchanges had existed in the earlier conflict, while the beginnings of exchange control in peacetime can be located in the 1930s. But it was not until the second world war that complete control was imposed on foreign exchange transactions, later to be made permanent by the Exchange Control Act of 1947.[28]

Though the control mechanism, as in the first world war, was built up piecemeal with much of it administered by private business itself, central direction was of course carried out by the Civil Service, including many new and temporary recruits from the private sector, on the basis of priority needs and budgetary programmes. It soon became clear on the basis of centralised programming of resources that the principal objectives must be those of securing the supply of arms and food and servicing the needs of the armed forces. Inevitably this meant the subordination of nearly everything else, that is exports and civilian production, since resource availability, even with the backing of lend-lease, was not sufficient to maintain prewar levels of activity in these sectors.

In fact, after the inauguration of lend-lease in 1941 the need to maintain export volumes to earn foreign exchange to pay for essential supplies was removed, and hence exports were no longer considered a priority. No doubt in time physical supply constraints would have dictated a diversion of resources away from exports, but probably not to the extent that actually occurred. By the middle of the war the volume of exports had slumped to less than one third of the prewar level and even by 1945 they were still less than one half. By that time only about two per cent of the labour force was producing for the export market as against about 10 per cent before the war. Imports too had been reduced sharply, though not by as much as exports, as a result of centralised buying and regulation, the main priority here being that of conserving shipping space.

It was on the civilian front however that control and allocation impinged most severely from the consumers' point of view, for it was from domestic resources that the main contribution had to come. All major items of food, except bread and potatoes, were rationed, while many unrationed foods such as vegetables and fruits were in short supply or unobtainable. For some items, for example fats, sugar and eggs, the ration allowances were minute, though overall an adequate dietary balance was maintained even if it was rather unappetizing, being heavily weighted with filler foods such as bread and potatoes. Rationing was also extended to non-food items such as clothing and furniture, backed by a scheme of utility products which were famous for their uniform drabness. The latter arrangements went hand in hand with a policy of concentration and contraction in what were regarded as less essential industries in order to release resources, especially labour, for more urgent tasks. Textiles, clothing and footwear were the most severely affected trades. Their total labour loss amounted to 648,000,[29] while one third of the cotton mills and more than 20 per cent of the wool textile mills were closed down.[30] The resulting cutback in production was substantial and stocks of clothing and footwear, both wholesale and retail, fell steadily in the last two years of the war. In fact in the case of clothing, stocks fell so rapidly that by the middle of 1944 they were below the level of efficient distribution. The supply of most other consumer goods, particularly furniture, furnishings and household products, was severely curtailed.[31] At the peak of the war effort it is probable that the production of consumer goods was little more than half that of prewar.[32] During the six years of war the civilian population had received less than four years' normal supply of clothing and less than three years' supply of household goods.[33]

Rationing, shortages and high taxation (see below) severely curtailed private consumption, thereby releasing resources for more pressing needs. By 1943 real consumers' expenditure *per capita* was over 20 per cent less than in 1938, though by the end of the war it had recovered somewhat from the trough. Even then, however, consumers' expenditure as a proportion of national income was little more than a half compared with 75 per cent before the war.[34] The sharpest expenditure reductions occurred in non-food items such as clothing, furnishings and household goods and private motoring, with falls of up to one half or more; the consumption of many basic foodstuffs was maintained quite well except for meat, sugar, tea and fats, while spending on beer and tobacco rose.

Unfortunately, from a longer-term point of view, the cutback in consumption was not matched by a commensurate decline in the real incomes of consumers despite heroic efforts in taxation. The main reason for this was that a loose wages policy coupled with extensive price control and subsidisation of basic foods allowed real earnings to rise. Wages were one of the few areas to escape control, largely because the trade unions, whose membership grew substantially during the war, rejected any form of centralised wage control which would interfere with their collective bargaining rights.[35] And in an effort to contain the wage-price spiral, stemming in part from the sharp rise in commodity prices at the beginning of the war, and on the grounds of social justice, the government committed itself to a policy of extensive price control and subsidisation of basic foods. The cost of these rose from £72 million in 1940 to £250 million by 1945. The result was that the rise in the cost of living was moderated in the later years of war, rising by less than 50 per cent between 1938 and 1945. Basic necessities such as food, fuel and light, housing, clothing and transport, which made up the greater part of the typical household budget, rose on average by considerably less than retail prices in general, whereas the price of drink and tobacco and durable and non-durable household goods increased much more rapidly. However, average earnings, under the influence of labour shortages, extensive overtime working and strong union pressure, rose by some 78 per cent over the same period, giving a rise in real earnings of around one fifth.[36] The increase in taxation was not sufficient to offset the combined effect of the rise in real earnings and the cutback in consumption through rationing and shortages and hence savings rose sharply. This meant that purchasing power was being accumulated for future use (see below).

In short, the control mechanism was extensive and effective for immediate needs but by suppressing market forces it stored up a latent inflationary potential for later years. The apparatus of control was certainly more efficient and successful than that which had operated during the first world war. As Youngson notes, 'it was a remarkable achievement to multiply the size of the armed forces tenfold, keep the volume of civilian employment almost steady, reorganise the entire structure of production, reduce civilian consumption by over 16 per cent, and have the cost of living rise by less than 50 per cent.'[37]

## Financing the War Machine

At the time of increased rearmament in the later 1930s it had been recognised that Britain would not be able to sustain a long war without assistance from other countries. The level of armament spending then obtaining was already beginning to put severe pressure on the external account and the exchange rate as resources were directed away from exports and import requirements rose, and it would only be a matter of time before Britain's small reserves were exhausted.[38] Secondly, it was also appreciated that a long and costly war would require herculean efforts on the domestic financial front to ensure that the productive effort was not impeded through financial stringency, and at the same time to avoid the inflationary implications of the method of financing practised in the first world war.

In the event, finance did not become a stumbling block to the prosecution of the war effort despite some initial hesitancy in subordinating finance to military strategy. The initial efforts to boost exports to pay for hard currency imports were abandoned once American lend-lease aid began to flow, while the sale of foreign assets and the accumulation of sterling balances provided additional means of squaring the external account (see below). As far as the domestic costs of the war are concerned, these were met by a combination of taxation and borrowing, with just over one half the total cost being paid for out of taxation as against less than one third in the first world war.

It was originally assumed that much of the internal costs of the war could be met from taxation rather than by placing the burden on future generations by resorting to extensive borrowing. This approach had the additional advantage of containing domestic spending thereby easing the potential inflationary pressures stemming from rising commodity prices and wage levels. Consequently taxation levels were increased sharply at the beginning of the war. An excess profits tax of 60 per cent was introduced in 1939 and raised to 100 per cent in the following year. The standard rate of income tax was increased progressively from 7s 6d in the pound at the start of the war to 10s 0d in 1941, the reduced rate was raised from 5s to 6s 6d, surtax was increased, while earned income relief and personal allowances were lowered. Purchase tax at 16½ and 33⅓ per cent was imposed in 1940 on a wide range of items and the duties on drink, tobacco and entertainments were also raised.

It soon became clear however that even with these severe tax impositions the costs of the war would outstrip their revenue yield. In the fiscal year 1940-41 government expenditure showed an enormous quantum leap as the economy geared up for total war, and short of imposing penal rates of taxation which would impinge heavily on the poor and possibly reduce the incentive to work, there seemed to be little prospect of meeting the war costs or closing the inflationary gap by taxation alone.[39] In fact the early acceptance of a policy of extensive food subsidisation and the imposition of price control and rationing acknowledged the need to employ measures to contain the rise in the cost of living. The cost of the subsidies further increased the government's expenditure bill.

Thus attempts to balance the wartime budgets by taxation were abandoned in 1941. The proposal to meet half the costs by taxation was deemed to be the best that could be achieved and subsequent wartime budgets did not raise income tax above the 1941 level though there were some adjustments to indirect taxes.[40] Kingsley Wood's budget of April 1941, though it raised tax rates significantly, effectively marked a departure from conventional practice by recognising that the widening gap between expenditure and revenue would have to be met in future by mobilising the nation's savings. It is also usually credited as being the first Keynesian budget insofar as it set out estimates of national income and expenditure from which it was then possible to determine the inflationary gap remaining at existing levels of income and taxation.[41] The new procedure followed broadly the recommendations of Keynes who, in the previous year, had published a pamphlet on *How to Pay for the War,* containing estimates of current national income as a basis for estimating the wartime inflationary gap which would need to be filled by savings once taxation had reached the acceptable limits. He also proposed the device of postwar credits, or deferred tax credits, to soften the burden of taxation, which was adopted in 1941.

For the remainder of the war therefore, the main preoccupation of fiscal policy was that of siphoning off the savings of the community to enable the government to pay its way. Borrowing constituted approximately one half of budgetary spending on average and most of it was raised from internal sources. The operation was remarkably successful in that by skilful manipulations of the money market the government was able to borrow much more cheaply than had been the case during the first world war. Apart from a brief spell at the start of the war, Bank rate was held at two per cent for the duration of war, which enabled the authorities to soak up most of the available savings at a low and steady cost to the Exchequer. Two principal means were used to channel savings into the government coffers. First, a strong patriotic appeal was made through the National Savings Movement to encourage small savers to invest their surplus funds in the Post Office, in the Trustee Savings Banks or in National Savings Certificates or Defence Bonds, with National War Bonds available for wealthier people. In addition, a Tax Reserve Certificate was introduced in 1941 to allow firms to invest for accrued tax liabilities. Any funds not invested through these media tended to go to the banks and these were then mopped up by government debt instruments including Ways and Means Advances, Treasury Bills and Treasury Deposit Receipts the last of these being a compulsory form of lending by the banks to the Exchequer. The most widely used instrument was the Treasury Bills which were absorbed in large

*Table 5: Exchequer Receipts, 1937-38 to 1944-45 (£mn)*

| Year ending 31 March | Taxes and duties | Debt creation | Misc. items | Total | (2) as % of (4) |
|---|---|---|---|---|---|
| | (1) | (2) | (3) | (4) | |
| 1937-38 | 948.7 | 265.9 | 1.6 | 1216.2 | 21.9 |
| 1938-39 | 1006.2 | 176.1 | 14.1 | 1196.4 | 14.7 |
| 1939-40 | 1132.2 | 800.6 | 8.2 | 1941.0 | 41.2 |
| 1940-41 | 1495.3 | 2491.8 | 26.3 | 4013.5 | 62.1 |
| 1941-42 | 2174.6 | 2696.6 | 51.5 | 4922.7 | 54.8 |
| 1942-43 | 2922.4 | 2814.6 | 38.0 | 5775.0 | 48.7 |
| 1943-44 | 3149.2 | 2759.1 | 38.4 | 5946.7 | 46.4 |
| 1944-45 | 3354.7 | 2827.0 | 39.5 | 6221.2 | 45.4 |

*Source:* Central Statistical Office, *Statistical Digest of the War*, p. 194.

quantities by banks, firms, financial institutions and the public at large. By the middle of 1945 Treasury Bills outstanding amounted to nearly £4 billion as against just over £2 billion in Treasury Deposit Receipts by banks, and the two together accounted for over 90 per cent of the floating debt of some £6½ billion (July 1945).[42] Table 5 indicates that borrowing rose to a peak of well over half total revenue by 1941-42 after which its share declined as tax revenue yields rose. Over 80 per cent of the revenue from all sources was spent on defence purposes from 1940-41 onwards.

On balance therefore, the domestic financing of the war was remarkably successful. The government managed to cover its outgoings without resorting to penal rates of taxation, by means of mobilising the nation's savings at low rates of interest. That it was able to borrow so heavily on such favourable terms was partly because of the tight restrictions on spending of both companies and individuals who were forced to accumulate their excess income which then found its way into government debt instruments. This accumulation of savings was to pose a problem for the future but for immediate purposes it served the country well. At the same time taxation, forced savings, subsidies and rationing curtailed personal spending by some 20 per cent in real terms and almost eliminated the inflationary pressure for the time being. After a sharp jump in retail prices of nearly a third between 1939-42, there was only a very modest rise during the remainder of the war. Of course, this merely served to bank up a latent inflationary potential for peacetime, but that was scarcely taken into the reckoning at a time of national emergency.

At the onset the external financing of the war presented an even bigger problem. Even before the war began Britain's balance of payments was showing the strain of the accelerated rearmament programme, and once hostilities commenced it was expected that it would become very much more acute. As domestic industry was diverted to military production the former peacetime level of exports could not be maintained, while rising world commodity prices and increased demand for imports would raise the total import bill. Invisible earnings were also likely to decline, while sterling would come under strain as capital outflows increased. In the event Britain's balance of payments on current account recorded large deficits in every year of the war, which, on the basis of revised figures, averaged £751.7 million for the six years 1940-45, compared with £55 million in 1938 and £250 million in 1939.[43]

Since Britain's holdings of gold and dollar reserves were so small — £450 million at the start of the war — there were no prospects of meeting the external deficits by drawing on these. Initially therefore, until longer-term solutions could be developed, Britain took steps to boost exports and control imports and foreign-exchange use. Early in 1940 a major export drive was started with particular emphasis on hard currency (dollar) areas, while controls were imposed on imports and exchange dealings to prevent a drain of currency abroad and to maintain the pound sterling at $4.03, that is some 20 per cent below its old gold parity. However, it soon became apparent that such policies were self-defeating. There was a limit to which imports could be cut given the need for essential supplies of food, materials and equipment, while the attempt to divert more resources into exports simply meant that less were available for essential war production. In fact by the end of 1940 it was already clear that Britain could not hope to pay for all her requirements, especially those originating in the dollar area, by normal peacetime trading methods.

Fortunately, the dollar problem was solved rather sooner than expected, in the form of lend-lease supplies from the United States. During the first year or so of the war America had been perfectly willing to supply goods to Britain so long as this country could pay for them, but the United States had no desire to provide outright assistance which would have involved her in abandoning a neutral stance and entering the war directly. But during the course of 1940 American attitudes changed when it was realised that Britain was rapidly running out of reserves and therefore could not possibly be expected to defend the freedom of Europe single-handed. At the end of that year therefore, President Roosevelt announced a programme of lend-lease aid, which was ratified by the US Congress in the following year. It provided for goods and services to be made available free of charge to Britain and her Allies for the duration of the war. However, the aid was also accompanied by a number of restrictive provisions. Britain was required to divest herself of all remaining holdings of dollars and to sell off important capital assets in the United States, in some cases at unrealistically low prices. Furthermore, lend-lease supplies were not to be used in British exports, while Britain had to promise not to discriminate in international trade after the war. Finally, when Britain's reserves began to increase in the summer of 1943 following American troop expenditure in this country, the United States began to scale down lend-lease allocations and to demand increased reciprocal aid from Britain.

In the difficult circumstances of the time Britain had little option but to accept these rather stringent conditions, and indeed it would no doubt have seemed ungracious to object to them seeing that the aid was being provided free of charge. At first lend-lease assistance only made a marginal contribution to the Allied war effort, but by the peak of 1943 and 1944 it was providing one quarter of the armaments received by British and Empire forces. In addition, lend-lease supplies of food and raw materials were also of crucial importance. Altogether some 85 per cent of all lend-lease supplies went to the British Empire and Russia.[44] Britain alone received lend-lease aid, including Canadian contributions, totalling £7,500 million, which, after allowing for reciprocal aid from the British side, left a net figure of £5,400 million.[45] Lend-lease certainly solved the immediate dollar problem and allowed Britain to divert resources away from exports. In fact by 1943 their volume had fallen to less than one third of the 1938 level and most of them went to areas capable of supplying Britain with strategic goods.

The second main solution to the external payments problem was the accumulation of sterling balances by countries in credit with Britain. In effect Britain paid for overseas supplies and the surrender of dollar earnings by members of the sterling area by building up postwar claims against herself in the form of blocked accounts in London. By the middle of 1945 these balances amounted to no less than £3,355 million, the greater part of which, £2,723 million, were on account of sterling area holders. India alone had balances of £1,321 million. Substantial sums were also held on behalf of several Colonial and Dominion countries and the Middle East, with the remainder being held by non-sterling countries including Argentina, Brazil and the West European governments in exile. As with lend-lease, they allowed Britain to draw upon a greater volume of resources than she would otherwise have been capable of paying for immediately. But unlike lend-lease they presented a burden for the future, not the least of which was the annual servicing charge incurred and the need to provide for the contingency of their eventual withdrawal.

The third major source of meeting our external commitments was the sale of assets held abroad. Over one quarter of Britain's prewar foreign investments of some £4 billion were liquidated, realising a total of around £1,100 million. The main disposals were in India and the United States, some of the latter, as in the case of Courtauld's Viscose Corporation, being made at a heavy sacrifice at the time when, before lend-lease, the US Administration 'were fertile of suggestions to the British for stripping themselves bare'.[46] The sales were of course to reduce Britain's future earnings potential on invisible account.

The final balance sheet on the external side can therefore be summarised as follows. Britain's total debits, that is payments required for imports of goods and services together with expenditure abroad on British troops and free deliveries from the dollar area, amounted to some £16,900 million from the outbreak of war until the end of 1945. Only £6,900 million of this total was met by earnings from the export of goods and services from Britain; the remaining £10,000 million was covered by net aid from North America of £5,400 million, sales of foreign investments of £1,100 million and the accumulation of debts, mainly sterling balances, of approximately £3,500 million.[47] But if Britain had solved the problem of financing the external costs of the war with relative ease, this was at no small cost for the future. Lend-lease aid did not present a difficulty since at the final tally Britain had only to pay a small amount for goods still in the pipeline following the suspension of aid on the conclusion of the war with Japan. Rather the main burdens arose from the inevitable neglect of export markets during the war, the loss of invisible earnings partly through dis-investment abroad, and the overhang of the sterling balances. These were not insuperable burdens by any means, but they were obviously going to require a substantial effort on the export front in the reconstruction period.

## War End Balance Sheet and the Tasks of Reconstruction

It is now time to take stock of the consequences of the war on Britain's economy and to determine the priorities for the reconstruction period. There has been in the past a tendency to overstress the damage and economic losses suffered by this country. These, it is true, were far from being negligible but when they are compared to what happened in many European countries they somewhat pale into insignificance. Moreover, in terms of future potential one should also take into account the accumulated stock of skills, expertise and resources which Britain still possessed.

In aggregate terms one cannot argue that Britain's output potential was devastated by the war. As we have seen, total output in real terms rose to a peak in 1943 after which it declined somewhat, though still remaining above the prewar level at the conclusion of hostilities. Of course, the structure of the product mix had changed considerably over the course of the war. Resources had been diverted from civilian production and exports to military goods on a vast scale so that by the end of hostilities the distribution of factors of production throughout the economy was in no way commensurate with the huge task of reconstruction to meet normal peacetime needs. Some one half of the labour force was in the services or munitions industries,[48] which meant that many civilian industries had seen their workforces seriously depleted. One of the worst affected was building and civil engineering where the workforce was less than half that of prewar, and a shortfall of labour was common to most industries producing for the domestic market or for export. Not surprisingly therefore, the output of most consumer goods was at a low level by the end of the war and private stocks were very much reduced.

It was expected, however, that the redirection of manpower towards peacetime tasks would be effected fairly quickly and smoothly and that it would not give rise to any severe or permanent costs. Manpower losses through death had been relatively light compared with the first world war — around 300,000 military casualties and a small number of civilians, though if one includes the wounded and incapacitated the total figure is nearer three quarters of a million. However, it was not anticipated that a setback of this magnitude would give rise to severe labour shortages and there was no question of the 'lost generation' of the first world war which had been much stressed at that time. Skill deficiencies might pose a problem insofar as those in service had lacked the opportunities for industrial training. There were also some exceptional and rather intractable manpower difficulties looming on the horizon. One or two of the basic industries had suffered badly during the war. Coal was obviously a case in point. By 1945 it was clear that cheap and plentiful coal was a thing of the past.[49] Efforts to rebuild the labour force in mining and to improve its efficiency had met with little success with the result that coal output and labour productivity had fallen steadily during the war. Clearly there were going to be problems on both counts in the future, though in part these problems predated the war. The textile industries, so important still to Britain's export trade, also faced a serious manpower problem since the labour force had been cut by over one third and early attempts to restore it had not met with a great deal of success. The construction industry would also need to rebuild its workforce.

Shortages of manpower in some sectors were by no means the only problem. The war had affected efficiency adversely in many cases, partly through a contraction in output and loss of key workers, but also on account of the neglect of investment and maintenance and repair. Both the coal industry and building recorded significant falls in labour productivity, as did several industries producing for the home market, such as textiles. On the other hand, there were compensations elsewhere. The great success story was undoubtedly agriculture with its increasing output and efficiency spurred on by rapid mechanisation and guaranteed markets. Many of the sectors producing munitions also gained in efficiency. In engineering, for example, the war had brought much that was new in mass production techniques some of which were readily transferable to peacetime activities, for instance motor manufacturing, and some which were not. The food industries, boots and shoes, tobacco and tinplate also recorded increasing returns. Overall there was no very strong trend movement in output per worker, which by 1945 was higher than in 1938 but less than at the peak in 1943, reflecting in part the decline in capital per worker after 1940[50]. On balance, manufacturing does not appear to have suffered long-term permanent damage to its efficiency.

In some respects the war had been advantageous to Britain's productive structure in that it boosted the potential growth industries and pulled the country away from its previous overcommitment to the ex-growth staples. The industries most favoured were aircraft production, engineering, chemicals, scientific instruments and pharmaceuticals, most of which had a bright future in contrast to sectors such as textiles and shipbuilding. Moreover, many of these growth industries were science-based and gained considerably from the more intensive application of science and technology stemming from wartime needs. Scientific discoveries and their application during wartime abound — magnetron valves, radar, jet engines and other aircraft developments, nuclear fission, antibiotics, DDT and other insecticides, electronic computing and control systems, artificial rubber, nylon and plastics.[51] If Britain failed to take full commercial advantage of these developments in peacetime it was not for want of ingenuity and resourcefulness on the part of her intellectual manpower.

A serious setback from the long-term point of view was the loss and destruction of Britain's physical assets which was on a greater scale than during the first world war. Some of the losses were quite formidable and owed much to the extensive aerial warfare. British shipping, for example, suffered heavy damage losing some 60 per cent of the prewar fleet, and even with replacements it was still 30 per cent smaller in June 1945 than it had been at the beginning of the war. Nevertheless, this was somewhat better than most other belligerents apart of course from the United States which increased its fleet holdings enormously in this period.[52] Even more visible to the general public was the impact on housing. The total number of houses which were damaged or destroyed amounted to over 4 million (about one eighth were destroyed or rendered uninhabitable), equivalent to about one third of prewar stock, to which must be added the damage and destruction to factories, plant and machinery, schools, hospitals and other buildings.[53] Apart from direct loss or damage to physical assets, one must also take account of the deterioration in capital stock through neglect of maintenance and renewal because of the allocation of resources to more pressing needs during wartime. Serious deficiencies developed in this regard in some of the basic industries, especially transport and coal, though in the latter case there was an increase in mechanisation, as too in the less essential industries catering mainly for the civilian market. The opposite was true of course in war-related industries: metals, engineering, chemicals and also agriculture, where significant increases in capital equipment were recorded. Overall, investment throughout the war declined quite significantly, especially in buildings, while after 1940 net additions to the capital stock were negative. Approximate estimates suggest a loss of about 10 per cent of prewar national wealth due to physical destruction and dis-investment, rising to 25 per cent if external dis-investment is included.[54] But most of the losses on the domestic side had been made good by 1948 except in mining, transport and the distributive services, and often with better vintage equipment, so that there is no case for arguing that Britain suffered a long-term permanent setback in this respect.

The general opinion at the time was that the most serious damage had occurred on the financial side, that is pertaining to Britain's external account. Lord Keynes described it as Britain's financial Dunkirk, no doubt anticipating the consequences of the impending termination of lend-lease. It is true that there were no claims for payment for Mutual Aid supplied during wartime, only for those goods still in the pipeline after the peace with Japan, for which Britain was required to pay $650 million. But in the interim reconstruction period Britain would desperately need further credits, a fact recognised by the despatch of a mission headed by Lord Keynes in September 1945 to negotiate with the Americans for the renewal of aid.

The facts of the matter are straightforward enough. During the war Britain had been transformed from the world's largest creditor to the world's largest debtor. Wartime sales of overseas assets amounted to £1,100 million or one quarter of the prewar total, external liabilities had increased by over £3,000 million, and there had been a drop in the gold and dollar reserves. At the same time Britain's exchange earning capacity had been much depleted. Exports were quite inadequate to pay for even the most essential imports of food and raw materials; though they had recovered from their wartime low, they were still little more than 40 per cent of 1938 volume in the first nine months of 1945. Moreover, the geographical distribution of our exports was not consistent with the source of imports. At the end of the war the UK was drawing 42 per cent of its imports from the western hemisphere but selling only 14 per cent of its exports there. Dollar-earning capability would therefore be crucial in the reconstruction period. Nor was there likely to be much relief from invisible earnings in the immediate transition period since these had been diminished by asset sales, shipping and commercial losses and a commitment to heavy military spending overseas, especially in Germany during the reconstruction period.[55] Moreover, the worldwide shortage of food and raw materials was expected to push up prices and add to Britain's payments problems through a deterioration in the terms of trade.

It is not surprising therefore that the British negotiators in Washington painted a gloomy picture of Britain's balance of payments in the immediate future and even beyond. The deficit on current account for the first peacetime year (1946), even with rigid import control, was put at £750 million, with dwindling deficits thereafter, amounting to a cumulative total deficit of £1,250 million for the years 1946-50. It was estimated that at a minimum Britain would need to raise her exports to 75 per cent above the prewar level in order to restore the economy to a sustainable equilibrium which could persist without measures of restriction or other defensive weapons of a type with which it was hoped to dispense.[56] The balance of payments was seen therefore as the central problem of the economy, and at the core of this problem was the shortage of hard (dollar) currency. This would require priority to be given to promoting exports and getting the right imports.

The economic situation would not have been so bad had it not been for the large accumulation of purchasing power which the war had left behind. Potentially demand was far in excess of supply at the then current price level. Estimates suggest excess demand at the end of the war in the order of 25 per cent.[57] This was far higher than in 1918 because much of the inflationary potential had been suppressed rather than released as had been the case in the previous war. Even the heroic efforts in taxation were not sufficient to prevent a large accumulation of purchasing power in the hands of the public with unsatisfied claims on production. The need to borrow and the measures taken to contain inflation had in themselves added to the problem: forced government loans, low interest rates, a loose wages policy, rationing, subsidisation of basic foods and restrictions on the output of consumer goods, had helped to suppress inflation and make it a relatively cheap war, but at the expense of large amounts of purchasing power being stored up for future use. The rise in real earnings and the shortage of goods meant that in 1945 only 61 per cent of total private income was spent on consumption compared with 83 per cent in the last prewar year. Despite the fact that out of this surplus a much larger share was appropriated in taxation, private savings rose from 6 to 16 per cent of the total, and even higher proportions were recorded in the war years. In the six years 1940-45 net new private savings totalled £8,500 million, including the great increase in government securities issued to the public as well as deposits in savings and other banks. The accumulated liquid balances of private business firms alone amounted to over £4,000 million.[58] In view of the shortage of goods likely to arise in the immediate reconstruction period there was little prospect of physical controls being abandoned until at least a good part of this excess purchasing power had been eliminated. As one commentator put it: 'This huge inflationary potential constituted the basic rationale for continuation of some kind of economic controls into the postwar period.'[59]

## Planning for the Peace

The prospect of serious economic difficulties after the war was fully appreciated by those responsible for preparing for reconstruction. Planning for the peace began in fact at a very early date (August 1940) and it soon became a more or less continuous preoccupation of the Cabinet and government.[60] These deliberations led to a steady stream of papers embodying the criteria on which future economic and social policy should be based, though recognising that economic factors, especially the state of Britain's external finances, would place a severe constraint on what could be achieved. The plans as they emerged may be divided into three groups: (1) short-term arrangements relating to unscrambling the war machine; (2) longer-term issues especially those connected with social betterment; and (3) international questions.

On the first of these issues there was fairly general agreement that in the transitional period of demobilising the economy the mistakes perpetrated after 1918 should be avoided. An official committee reporting in 1942 discussed in some detail the events leading to the boom and bust mentality which had characterised the years 1919-21, and it was at pains to stress that the same sequence of events should not be repeated a second time. It pointed out that similar economic conditions, though in more acute form, would be operative at the end of hostilities, for example, more physical damage, greater accumulated purchasing power, a weaker external position etc., which would restrain the rate at which things could return to normal. The committee identified three top-priority tasks: the need to contain inflation; to transfer resources to the most essential jobs; and the urgency of dealing with the expected deficit in the balance of payments. The last of these was deemed the most crucial for if it could not be eliminated 'the whole fabric of reconstruction is in danger and we run the risk of a failure to maintain essential imports and of a major inflation.'[61]

Given these urgent tasks in a very tight resource situation, it became increasingly obvious that the complex network of controls built up in wartime could not be dismantled rapidly, as had been the case following the first world war and with disastrous consequences. Allocation and control of resources would still be required for some time to prevent the economy from short-circuiting under the pressure of the enormous volume of unsatisfied wants. The white paper on *Employment Policy* published in May 1944 (see below) recognised the need for continuing controls in order to establish certain broad priorities and to direct the efforts of industry towards the right tasks.[62] Some adaptation and modification of the control apparatus would no doubt be required to suit the more varied needs of peacetime since there would no longer be the single criterion of unlimited war needs to justify its operation, while there would be pressure to relax it coming from a variety of quarters. On the other hand, if the key objectives were to be achieved it would not be possible to abandon controls piecemeal: the overall framework of control would have to be kept intact since wartime experience had demonstrated beyond any shadow of doubt that to be successful the control system had to be comprehensive over the whole range of economic activities.[63] As it turned out, this appeared to fit neatly into the predilections of the Labour Government returned to office in 1945, who viewed the inheritance as 'a gift from the Gods'.[64]

At the same time considerable attention had been given from quite early on in the war to the details of unscrambling the war machine. Such matters included the liquidation of war contracts already occurring in some cases in the later stages of war, the sale of surplus government stocks, the disposal of government factories and the derequisitioning of factory space and shipping and of course the demobilisation of the armed forces. As regards the last of these, a far more satisfactory scheme of 'demobbing' was devised than that based on industrial category practised after 1918. The order of release from active duty was to be based on age and length of service as far as possible, though some preference would be given to key groups of workers required to carry out urgent reconstruction work. All were to receive gratuities, free clothing and paid settlement leave. Similarly, the disposal of government stores, the release of factory space and the derequisitioning of shipping were to be carried out in a more orderly manner than had been the case after the first world war.

At the same time numerous studies were made of individual industries outlining their needs, prospects and role in peacetime, and also emphasising the need to improve efficiency in many cases. Agriculture was accorded special attention in the later stages of war owing to increasing anxiety as to the availability of overseas supplies and the ability to finance imports in the transition period. Consequently, in 1944, the War Cabinet gave an assurance that the guaranteed food prices and assured market would be continued until the end of the 1947 harvest.

On the longer-term issues of social betterment and the reconstruction of infrastructure facilities including housing, much thought and planning had been in evidence throughout the war. One of the top priorities was housing, understandably in view of the considerable war damage to residential building and the fiasco following the first world war when the grandiose promise of 'homes fit for heroes' had failed to materialise. Various plans and projections were presented, the last being that of March 1945 which proposed the building of 300,000 permanent houses by the end of the second year following the termination of the war in Europe. This was a rather ambitious objective as it transpired, and many homeless were to find themselves housed in temporary 'prefabs' under an emergency holding programme until such time as more permanent dwellings could be built. Apart from residential accommodation, considerable attention was also given to the building or repair of schools, factories and infrastructure facilities including roads. In this context land use and development featured prominently and in 1943 an independent Ministry of Town and Country Planning was set up with responsibility for these matters.

Employment policy after the war also became a central issue, no doubt influenced by the unhappy experience of the interwar years. A paper on the maintenance of full employment in peacetime by the

Economic Section of the War Cabinet Offices was being circulated as early as 1941. In November of the following year Beveridge's famous report on social services was released, which had as its basic premise a high level of employment. The level of unemployment assumed in the report was unbelievably high however, though Beveridge later scaled it down to around 3 per cent in his own independent study.[65] During 1943 employment and related issues were thoroughly examined by a small group of officials and by May of the following year the government was in a position to issue a white paper on the subject, giving its blessing to the maintenance of a high and stable level of employment in peacetime.[66] This, it implied, could be achieved by regulating the demand for goods and services in such a way as to avoid large-scale unemployment, though it did stress that the success of this policy would depend on the stability of prices and wages and on a reasonable degree of mobility between occupations and regions. Severe structural unemployment imbalances, particularly geographic ones, would be dealt with through a more active regional policy.

Of all the schemes and plans debated during the period the one that caught the popular imagination most of all was that relating to the reform of the social services. These had grown up piecemeal during the twentieth century, and though by no means insignificant in coverage and depth, there were still glaring gaps in provision and a want of comprehensiveness. In 1941 therefore, the government commissioned Sir William Beveridge to carry out a comprehensive survey of existing schemes and facilities. His report appeared in November 1942[67] and immediately became a best seller, with 250,000 copies of the full version and 350,000 of an abridged one being sold within a few months of its appearance. Few official documents can have been so widely read.

The central objective outlined in the study was that of eliminating want and deprivation. To this end the report listed six main criteria for social-service provision and reform. In the first place this would require the creation of a comprehensive system of welfare covering the whole population from the cradle to the grave, and including unemployment benefit, sickness benefit, disability allowances, workmen's compensation, pensions (for old age, widows and orphans), child benefits, maternity and funeral grants, together with residual national assistance to cover those still falling outside the main schemes. In addition, it presupposed the setting up of a comprehensive national health service. The second principle was that of administrative unification in place of the confusing collection of bodies — government departments, local authorities and agencies — responsible for various social provision prior to the war. In essence this called for a single ministry and a single payment by employer and employee to cover all contingencies. The other four principles enunciated included the payment of adequate benefits, flat-rate benefits and flat-rate contributions, and a classification of the population into occupational groups with specified contribution rates and benefit entitlements.

There was, as might be expected, some reluctance on the part of the Coalition Government to accept these proposals in full largely on the grounds of the anticipated high cost of the whole deal. Nevertheless, the report did not want for official attention. It was studied in great detail during the next two years in the course of which the government modified many of the original proposals, especially that relating to adequacy. Eventually it published its own plans in two white papers in September 1944,[68] based largely on the main principles of the Beveridge study. Earlier in the year it also made known its scheme for a national health service, the main principle of which would be the provision of a comprehensive medical service for all persons free of charge.[69]

Meanwhile, there was an equally sustained effort to improve the climate of international economic relations in the postwar period, perhaps not surprising in view of the unsatisfactory experiences of the 1930s. Much of the initiative in this respect came from the two principal Allies, the United States and Britain, who organised a series of meetings to discuss a wide range of issues, including international finance, commercial policy, commodity problems, employment and many other matters. Several important international organisations emerged from these deliberations. One of the first was the establishment of an organisation, the United Nations Relief and Rehabilitation Administration (UNRRA) in November 1943, to supervise plans for short-term relief in war-devastated areas throughout the world once fighting had ceased. Britain as a founder member assumed an obligation amounting to some one per cent of her national income for such purposes, which was additional to the sums provided for the military administration of relief in the main occupied areas immediately following liberation. In the same year postwar food problems featured on the agenda of a series of Allied conferences on long-term problems of international policy at Hot Springs, from which eventually emerged the Food and Agricultural Organisation of the United Nations at the end of 1944, with Britain as a member. But perhaps the most significant development was in the field of international finance and banking, with the setting up of the International Monetary Fund to deal with international liquidity questions, and the International Bank for Reconstruction and Development for long-term loans, both of which were ratified at the famous Bretton Woods Conference in July 1944. These institutions, whose initial conception owed much to the work of Lord Keynes and Harry D. White, were to play a significant role in improving the state of international economic relations in the postwar era.

These are only a sample of the preparations being made for the future. Against the background of the pressing events of a total war, it is quite remarkable how much time and energy, and from a very early date, went into planning for the peace. Many of the longer-term plans, which could not readily be implemented by a coalition government, were no doubt ambitious in the light of the economic constraints obtaining after the war and would require modification in due course. But many of the policy changes, especially those relating to social welfare, which were to take place under the Labour Governments of 1945-51, were based on the preparatory work initiated during the war. The Labour Government no doubt moved in a more radical direction than had been intended by the coalitionists of wartime, but whichever government had been in power at the time there would almost certainly have been significant policy changes affecting the bulk of society. For in an important sense the planning of peacetime reconstruction reflected an implied contract between King and Country. 'There existed, so to speak, an implied contract between Government and people; the people refused none of the sacrifices that the Government demanded from them for the winning of the war; in return, they expected that the Government should show imagination and seriousness in preparing for the restoration and improvement of the nation's well-being when the war had been won. The plans for reconstruction were, therefore, a real part of the war effort.'[70] When the time came for the state to fulfil its part of the bargain the nation would be able to judge whether it had all been worthwhile.

## Notes

1. The preliminary general volume in the official civil series of wartime studies is that of W.K. Hancock and M.M. Gowing, *The War Economy* (1949).
2. For an interesting re-interpretation of the conventional Blitzkrieg view see R.J. Overy, 'Hitler's War and the German Economy: A Reinterpretation', *Economic History Review*, 35 (1982).
3. R.J. Overy, *The Nazi Economic Recovery 1932-1938* (1982), p. 51.

4. Hancock and Gowing, *op. cit.*, p. 71; see also B. H. Klen, *Germany's Economic Preparations for War* (1959).
5. G. C. Peden, 'A Matter of Timing: The Economic Background to British Foreign Policy, 1937-1939', *History*, 69 (1984), p. 25.
6. Hancock and Gowing, *op. cit.*, p. 549.
7. For employment data see C. H. Feinstein, *National Income, Expenditure and Output of the United Kingdom, 1855-1965* (1972), T126.
8. A. S. Milward, *War, Economy and Society, 1939-1945* (1977), p. 219.
9. D. H. Aldcroft, *British Railways in Transition* (1968), pp. 98-100.
10. C. B. A. Behrens, *Merchant Shipping and the Demands of War* (1955), p. 5, note 1.
11. M. F. W. Hemming, C. M. Miles and G. F. Ray, 'A Statistical Summary of the Extent of Import Control in the United Kingdom since the War', *Review of Economic Studies*, 26 (1959), p. 104.
12. Wages were one of the few exceptions owing to the opposition of the trade unions to any form of centralised wage control.
13. A. J. Youngson, *The British Economy, 1920-1957* (1960), pp. 141-2.
14. For wartime production see M. M. Postan, *British War Production* (1952); J. D. Scott and R. Hughes, *The Administration of War Production* (1955); J. Hurstfield, *The Control of Raw Materials* (1953).
15. The number of non-industrial civil servants rose from 387,700 in April 1939 to 704,700 in October 1945. J. Stevenson, *British Society 1914-1945* (1984), p.445.
16. Central Statistical Office, *Statistical Digest of the War* (1951), p. 55.
17. Feinstein, *op. cit.*, T52.
18. For food and agriculture in wartime see E. H. Whetham, *British Farming 1939-49* (1952); R. J. Hammond, *Food*, 2 Vols. (1951, 1956) and *Food and Agriculture in Britain in 1939-45: Aspects of Wartime Control* (1954).
19. *Statistical Digest of the War*, p. 60; S. Pollard, *The Development of the British Economy, 1914-1980* (1983), p. 205; K. G. Fenelon, *Britain's Food Supplies* (1952), pp. 182-3; B. W. Lewis, *British Planning and Nationalization* (1952), p. 259.
20. For agricultural policy in wartime see A. H. Murray, *Agriculture* (1955).
21. *Statistical Digest of the War*, p. 63.
22. Hancock and Gowing, *op. cit.*, p. 550.
23. The more equitable distribution of the available food supplies together with an improved dietary balance meant that a large part of the population was better fed by the end of the war than in peacetime. Nor did the population suffer from obesity.
24. Milward, *op. cit.*, p. 253.
25. See *Statistical Digest of the War*, p. 75 *passim*; Hancock and Gowing, *op. cit.*, Chapter 16; W. H. B. Court, *Coal* (1951).
26. Youngson, *op. cit.*, p. 152.
27. G. D. N. Worswick and P. H. Ady, *The British Economy, 1945-1950* (1952), p. 127.
28. R. S. Sayers, *Financial Policy, 1939-45* (1956), pp. 226-32.
29. *Ministry of Labour Gazette*, August 1945, p. 127.
30. E. L. Hargreaves and M. M. Gowing, *Civil Industry and Trade* (1952), pp. 364, 392.
31. *Ibid.*, pp. 475, 650, 652; E. Estorick, *Sir Stafford Cripps: A Biography* (1949), pp. 334-5.
32. S. E. Harris, *Economic Planning* (1949), p. 161.

33. *Economic Survey for 1947*, Cmd. 7046 (1947).
34. Feinstein, *op. cit.*, T42; J.C.R. Dow, *The Management of the British Economy, 1945-1960* (1964), p. 15.
35. P. Inman, *Labour in the Munitions Industries* (1957), pp. 316-18.
36. Feinstein, *op. cit.*, T134, T135, T140.
37. Youngson, *op. cit.*, p. 152.
38. G.C. Peden, 'A Matter of Timing: The Economic Background to British Foreign Policy, 1937-1939', *History*, 69 (1984), p. 17; R.A.C. Parker, 'The Pound Sterling, the American Treasury and British Preparations for War, 1938-9', *English Historical Review*, 98 (1983).
39. Hancock and Gowing, *op. cit.*, pp. 327-8.
40. For the wartime budgets see B.E.V. Sabine, *British Budgets in Peace and War, 1932-1945* (1970).
41. See R.S. Sayers, '1941 — the First Keynesian Budget', in C.H. Feinstein (ed.), *The Managed Economy: Essays in British Economic Policy and Performance since 1929* (1983).
42. *Statistical Digest of the War*, p. 196.
43. Feinstein, *op. cit.*, T82, T83.
44. R.F. Mikesell, *United States Economic Policy and International Relations* (1952), p. 91.
45. Feinstein, *op. cit.*, p. 113.
46. Hancock and Gowing, *op. cit.*, p. 233.
47. Pollard, *op. cit.*, p. 217.
48. *Ministry of Labour Gazette*, August 1945, p. 127.
49. Hancock and Gowing, *op. cit.*, p. 550.
50. Feinstein, *op. cit.*, T52.
51. Pollard, *op. cit.*, p. 202.
52. *Statistical Material presented during the Washington Conference*, Cmd 6707 (1945).
53. Worswick and Ady, *op. cit.*, p. 21.
54. Hancock and Gowing, *op. cit.*, p. 551.
55. For example, the interest each year on overseas investments, which prewar paid for 21 per cent of our imports, covered only 3 per cent by 1949.
56. Cmd. 6707 (1945), *op. cit.*, p. 5.
57. See J.C.R. Dow's memorandum to the *Radcliffe Committee on the Working of the Monetary System*, Memoranda of Evidence, Vol. 3 (1960), p. 95, para. 128.
58. T. Balogh, *The Dollar Crisis* (1949), p. 235; 'Three Aspects of Transition', *Midland Bank Review*, May 1946, pp. 2-3.
59. N. Rosenberg, *Economic Planning in the British Building Industry 1945-49* (1960), p. 15.
60. Sayers, *op. cit.*, p. 98.
61. Hancock and Gowing, *op. cit.*, pp. 535-6.
62. *Employment Policy*, Cmd. 6527 (1944), para. 18.
63. Hancock and Gowing, *op. cit.*, p. 536.
64. J. Jewkes, *Ordeal by Planning* (1948), pp. 76-7.
65. W.H. Beveridge, *Full Employment in a Free Society* (1944).
66. *Employment Policy*, Cmd. 6527 (1944).
67. *Report on Social Insurance and Allied Services*, Cmd. 6404 (1942).
68. *Social Insurance*, Parts I and II, Cmd. 6550 (1944), Cmd. 6551 (1944).
69. *A National Health Service*, Cmd. 6502, (1944).
70. Hancock and Gowing, *op. cit.*, p. 541.

# 8 Facing the Future with Labour, 1945-1951

## New Directions

When hostilities in Europe ceased on 8 May 1945 the wartime Coalition Government headed by Winston Churchill rapidly disintegrated. At first there was some hope that it might remain in office until the conclusion of peace with Japan, but less than a fortnight later Attlee and his Labour ministers resigned, leaving Churchill in charge of a Caretaker Government consisting mainly of Conservatives. An election was therefore called immediately, to be held on 6 June, with the results delayed until 26 July because of the time required to collect the servicemen's vote.

When the returns were finally announced the Labour Party was literally spellbound; they captured no less than 393 seats as against only 210 for the Conservatives and their allies, while there was a sprinkling of 12 Liberals and 23 Others. Labour secured 48 per cent of the total votes cast compared with less than 40 per cent by the Conservatives, and they made spectacular gains in the major cities, in many southern dormitory suburbs and even in some of the rural constituences.[1] It was undoubtedly Labour's 'finest hour'. For the first time Labour had a clear and overwhelming majority which gave it a mandate to redress the missed opportunities of the first two minority Labour governments of 1924 and 1929-31.

The somewhat unexpected result has been interpreted in various ways by political historians. There is little doubt, however, that economic and social factors played a significant part in the sweeping victory of the Left. Labour's election manifesto, *Let Us Face the Future*, caught the mood of the time simply because it appeared to offer the prospect of something better than a return to the mass unemployment and inegalitarian society of the interwar years. The people had no wish to return to the 'normalcy' of the prewar world with all its unpleasant associations, as had been the case after the first world war when the gradiose reconstruction promises of the

Lloyd George Coalition Government had been broken. Thus, while the nation may have been galvanised into expressing its political preference in response to the experience of the war years, the election result was not simply a challenge to the Coalition Government of 1940-45, but rather a response to 'the total record of all the governments of the twenty years between the wars'.[2] There was for the first time, it appeared, a Brave New World in prospect and Labour appeared ideally suited to drag the nation into it, rather than pulling it back into the Cruel World of the past.

Whatever the circumstances surrounding the victory of the Left, it was soon apparent there were going to be great changes in the direction of policy compared with the past. Ideological substance committed the Party to certain principles, particularly with regard to economic and social matters, and its strong electoral showing clearly presented the opportunity to put these into practice to a degree not possible under previous Labour governments. These policies might have to be modified in practice by the force of events and to conform to the basis of a new moderate consensus of the Left, but they owed much to the inherited legacy of a Party whose ideals had so far been frustrated. On the other hand, one can easily overemphasise this point. It should not be forgotten that in many respects Labour's policies, particularly those on employment and welfare, were but an extension of the reconstruction blueprints drafted during the wartime Coalition Government.

There is little doubt that Labour assumed power committed to a substantial programme of economic and social reform. One such reform was public ownership and during the first year of office nationalisation measures were rushed through the legislature with almost indecent haste. The issue of public ownership had long featured on the agenda of Labour Party conferences. *Labour's Immediate Programme* of 1937, for example, had called for a wide measure of public ownership together with an extension of public intervention in economic affairs, including control over the levers of finance.[3] These aims were reaffirmed during the war and became part of the 1945 election manifesto. Prime targets for national ownership were the Bank of England, the coal industry, electricity and gas undertakings, the railways together with other forms of transport such as civil aviation, and iron and steel, the latter in response to a resolution carried at the 1944 Party Conference. Land nationalisation was however quietly dropped from the list in favour of the control of land use and development, while no specific mention was made of the joint stock banks.

If nationalisation became a central issue in the Labour programme as an antidote to what was termed the discredited capitalist system of the past, there was far less clarity about the mechanism by which public control was to be effected or about the long-term objectives of carrying out such a programme. In matters of detail such as organisational structure, the integration of services, finance, pricing, compensation to stockholders and the relations between the industries acquired and the consumers and employees, very little serious thought had been given. Nor was there any greater clarity about the real purpose and benefits of nationalisation. It is true that socialisation of the means of production was regarded in a general way as offering a better prospect for both consumer and worker, but beyond that the matter was left entirely vague and there was no indication of how or whether public ownership would be beneficial from the point of view of Britain's postwar economic problems. Some of the earlier proposals on nationalisation, especially of the basic industries, had stressed the need for state ownership on social grounds, including wealth distribution and improving industrial relations. But by 1945 the emphasis had shifted more to economic considerations: in the case of the old industries such as coal and the railways it was a matter of improving efficiency, whereas for iron and steel, civil aviation and Cable and Wireless, it was a question of integrated planning, investment and re-equipment. Yet given the limited and selected nature of the shopping list it is difficult to see how the second objective could be achieved with any real measure of success. Indeed, there was no systematic coherence about Labour's public ownership programme from the point of view of the economy as a whole. In fact in later years of the Administration, when some disillusionment crept in as to the merits of public ownership and the wisdom of further extensions at the time of the proposed legislation for steel, the relevance of state ownership on national economic grounds was called into question. Lincoln Evans, the General Secretary of the iron and steel workers union, whose membership were not particularly enthusiastic supporters of nationalisation of their industry, expressed considerable scepticism about the merits of further state holdings. In a letter to Morgan Phillips (2 October 1950), the General Secretary of the Labour Party, he wrote: 'Public ownership schemes of this kind were no solution to the problem of our present economic difficulties, and had very little relevance to such matters as our trading prospects in the world, the balance of payments, or, indeed, to making ourselves economically independent by 1952.'[4]

Labour's enthusiasm for planning and controls also waned as time wore on. Labour could not have come to power at a better time given its espousal of more intervention on the economic front and its professed predilection for using planning and controls to attain its socialist objectives. It inherited all the machinery of wartime planning and a battery of economic controls. In its election manifesto its commitment to planning appeared to be total: it would 'plan from the ground up, giving an appropriate place to constructive enterprise and private endeavour in the national plan ... [with] a firm constructive hand on our whole productive machinery.' Controls would be an integral part of the planning process: 'It is either sound economic controls or smash.'[5] Attlee's debate on the Address in 1945 confirmed the substance of the election promise: 'it would be absolute madness to abandon those financial and economic controls which have served us well during the war.'[6] Even as late as June 1949, by which time planning and controls were beginning to lose much of their attraction, Herbert Morrison still maintained that, 'We do not believe that we can constructively develop and maintain a high degree of productivity and a considerable and adequate amount of production unless we have economic planning and control on behalf of the community.'[7]

It is not difficult to see why the Labour Party should have embraced planning and controls with such fervour. They fitted in well with the Party's basic philosophy on economic policy and appeared to offer a marvellous opportunity for the government to carry out its major economic and social reforms. Planning and controls had worked well in wartime and many of the benefits flowing from them were readily identified in the public's mind with Labour ministers.[8] From a more practical point of view planning and controls would avoid the nasty effects of market forces, a key consideration in the immediate postwar period when all resources were scarce. Furthermore, direct intervention predated Keynesian economic analysis and since the importance of the Keynesian revolution was not then fully appreciated by the Labour Party it seemed safer to rely on the more familiar types of control, for the time being at least.[9]

As things turned out the Labour Government did not accomplish a great deal on the planning front — in fact they proved somewhat inept when put to the test. The system of controls was retained for some time though as conditions improved it began to crumble. As controls were abandoned the role of budgetary policy in the overall management of the economy increased, though monetary policy was not to emerge as a major policy instrument until after the return of the Conservatives to power late in 1951.

Another main plank of Labour's progamme was of course social reform. Again this subject had long been one of the basic issues of the Party's agenda and the poverty and unemployment of the 1930s gave added zest to its reformist zeal in this direction. However, the first major practical thrust towards comprehensive social reform came with the series of studies carried out during the war, especially the report by Beveridge which called for a system of social security from the cradle to the grave which would banish the principal wants of society. The Labour manifesto of 1945 gave special prominence to social reform covering a wide field. It reaffirmed the key sections of the Beveridge proposals and enlarged upon them, with particular emphasis on a national health service, full employment and a large building programme. Above all, the system of social security would be comprehensive rather than patchwork as had been the case hitherto, whatever the cost to the public purse.

Thus, apart from the immediate tasks of reconstruction and demobilisation and the need to expand production and exports, the Labour Government had an impressive policy programme which it was determined to implement come what may. It wasted no time in making a start on its large legislative agenda. In the first full year of office it had already pushed through several nationalisation Bills and social security measures. For the most part it managed to implement the major part of its long-term programme though at some cost to immediate welfare. Tight constraint had to be maintained on consumption and in some years conditions proved more austere than during wartime. But the overriding constraint throughout the reconstruction period was Britain's weak external position and the periodic strains on sterling. It was this problem that the new Labour Government had to face soon after taking office and it is appropriate therefore to turn to this matter first.

## Sterling under Pressure

One of the dominating features of the postwar period was the shortage of freely convertible currency both in Britain and Europe. Britain emerged from the war as the world's leading debtor country with her export potential and earnings capacity badly depleted, so that it was scarcely surprising that the external account should present problems. However, the payments problem was very much a dollar one since in most years Britain managed to earn a surplus on trading with non-dollar areas of account. Thus on average the dollar deficit accounted for more than the total deficit on current transactions throughout the years 1946-51, and even in years of overall surplus, 1948 and 1950, Britain recorded substantial imbalances with the dollar area. The reasons for this persistent dollar problem can be readily identified. Britain's export and invisible earnings in the dollar area had shrunk dramatically during the war, and it took time to restore them especially since the United States had become even more self-sufficient in the meantime. Moreover, in the interim reconstruction period Britain was more dependent than formerly on dollar imports since the dollar area was virtually the only source from which imports of food, raw materals and equipment could be readily obtained. The same could be said of many other countries, especially European, and the resulting scramble for commodities led to price rises thus exacerbating the dollar deficits through deterioration in the terms of trade. Finally, Britain found it difficult to earn dollars in third-world countries since dollar currency was at a premium worldwide.[10]

As noted in the previous chapter, ministers and Treasury officials were fully aware of the impending postwar weakness on the external account. What they failed to appreciate fully was the severity of the dollar problem and the speed with which it in fact emerged. Labour had no sooner taken office than the Japanese surrendered, whereupon President Truman announced the abrupt termination of lend-lease (August 1945). This was a severe blow to Britain since it had been anticipated that American aid would continue at least until she had been able to make a start in converting her economy to a peacetime basis which would ease the task of financing import requirements. The immediate prospects were therefore extremely bleak unless Britain could secure additional credit to tide her over the difficult transition period. Keynes and Lord Brand were hastily despatched to the United States with instructions to negotiate a new financial package. Their efforts produced the famous Dollar Loan Agreement of December 1945 which was to become operative in July of the following year. It was far less generous than Britain had hoped for, but force of circumstances compelled the government to accept it. The main provisions were as follows: the USA offered Britain a line of credit of $4,400 million of which $650 million was for the purpose of clearing outstanding debts under lend-lease pipeline goods coming in after VJ Day. Canada also chipped in $1,250 million giving a net credit of $5 billion. The terms of the loan were generous, interest at 2 per cent payable with capital redemption over a 50-year period starting in 1951. But the sting was attached in the additional provisions. One was an undertaking to reduce the sterling balances which had accumulated during the war; this in the long-term would be advantageous to Britain but short-term would present problems. Far more crucial in the light of subsequent events was the obligation to apply the Bretton Woods agreements much earlier than originally intended, namely making sterling convertible and removing discriminatory trading practices. These conditions were to apply one year after signing the agreement, that is by July 1947.[11]

This new line of credit was intended to last Britain until at least 1951, by which time it was expected that some semblance of equilibrium would have returned to the balance of payments. The projected deficits for 1946 and 1947 were large and it was fully expected that much of the loan would be used to cover these, whereas in later years the rising volume of exports and invisible earnings would lessen the need for assistance of this kind. Unfortunately, things did not quite turn out as planned; the loan was used up rather more rapidly than originally envisaged partly because of a combination of unfavourable circumstances. This meant that by the time the date for the convertibility of sterling approached (15 July 1947) the line of credit was nearing exhaustion.

The sterling crisis of the summer of 1947 needs to be set against the context of preceding developments. When the Loan Agreement came into force in July 1946, progress was already being made in converting the economy to a peacetime basis, output was recovering and exports for the year as a whole were almost back to the prewar level. The latter was quite a remarkable achievement given the very low level of exports in 1944, but one should bear in mind that it was a seller's market worldwide at this time and Britain's production potential was in better shape than that of many other countries. Moreover, since imports were tightly controlled, the balance of payments shortfall, and even the dollar deficit, was less than anticipated, and in any case the US loan could take care of any problems on that score.

Further progress was anticipated in 1947 but almost from the start of the year things began to go wrong. Dollar imports were rising very rapidly, partly on the strength of the loan but also because of lax domestic policy in the year or so following the end of the war which led to a sharp rise in consumption. Exports, on the other hand, began to stagnate due to supply difficulties at home. Between July 1946 and June 1947 imports from the dollar area rose to $1.54 billion while exports to that area only amounted to $340 million. On the domestic side, the sharp rise in consumption between 1945 and early 1947 diverted resources away from exports, while the severe weather of the winter of 1946-47 and the accompanying coal shortage and power cuts reduced the export potential further, probably acounting for an export loss of some £200 million. Thirdly, the terms of trade continued to deteriorate due to the world scarcity of dollars. Fourthly, the costs of Britain's overseas military commitments were increasing especially in Germany and the eastern Mediterranean. In 1947 for example, military spending abroad was no less than £209 million which more than consumed the total invisible earnings from foreign dividends, interest and profits, whereas between 1934-38 government spending abroad on military, administrative and diplomatic account was only £6 million a year.[12]

But the deterioration in Britain's current balance between 1946 and 1947 can only account for part of the enormous loss in reserves during the first three quarters of 1947.[13] More important was the capital drain due to the looseness of the arrangements which had been made to tie up the sterling balances. The overseas sterling area ran into deficit during 1947 as against a small surplus the previous year, and because of inadequate exchange control sterling countries were drawing on the common dollar pool to meet their current requirements.[14] In addition, after February 1947 the position was exacerbated by the loopholes in the transferable account system. By joining the transferable account area European countries could earn convertible sterling before July. By restricting imports from Britain and increasing their exports they were able to enlarge their surpluses with this country and thereby transfer their dollar problem to Britain.[15]

The combination of these adverse factors meant that even before the appointed day for convertibility (15 July) the drain on the dollar credits was considerable. By the end of June 1947 undrawn credits were down to $2,250 million, or 45 per cent of the original total. Once convertibility became *de jure* the loss accelerated sharply as holders of sterling took the opportunity to obtain scarce dollars. Within the space of a few weeks about one half the remaining credits had disappeared and on 20 August Britain was forced to abandon the short-lived experiment in convertibility.

Had convertibility not been suspended it is probable that most of the remaining credits would have been exhausted by early October at the latest, and at that point Britain would have had to draw on her small gold and dollar reserves. Once these had gone she would have had no option but to impose very severe cuts in imports and domestic activity in order to balance the external account. As it was, the government was forced to implement import cuts in the late summer and autumn of 1947, as did other sterling-area countries, together with cuts in projected investment and a restrictive budget in November to curtail domestic consumption.

Though these measures helped to contain the dollar drain to manageable proportions, it was clear that without a substantial underlying improvement in exports, or unless further aid was forthcoming, Britain had little prospect of relaxing the tight policy stance on imports and domestic consumption since the reserves were too small to stand any persistent strain.

The dollar crisis of 1947 was not solely a British phenomenon. In fact it affected most of Western Europe, threatening recovery and putting an end to the American dream of multilateral non-discriminatory trading and exchange convertibility. Ambitious plans for economic expansion and social reform both in Britain and Europe were also in danger of being curtailed unless the persistent dollar shortage could be eased in some way. It was against this background and the emergent cold war between East and West that General Marshall announced a massive programme of aid for Europe in the summer of 1947, which was approved by Congress in the following Spring. The aid was made conditional on Europe adopting a measure of economic cooperation and it was accepted readily by European governments whose representatives met in September 1947 as the Committee of European Economic Co-operation (CEEC) to draw up a plan for its distribution over a four-year period.[16] The aid began to flow from April 1948 and the total disbursed through to the end of June 1951, when what was left of the European Recovery Programme was merged with the US defence support programme, amounted to $12,000 million.[17]

Under Marshall aid Britain received $2,694 million between April 1948 and the end of 1950, that is 22 per cent of the total. Most of this allotment had been taken up by the middle of 1950 and Britain was the first country to suspend Marshall aid and stand on her own feet. At the peak (1949) Marshall receipts accounted for 2½ per cent of Britain's national product and 12 per cent of her total imports, while they covered no less than 57 per cent of her dollar imports including large fractions of essential foods and raw materials — four fifths in the case of aluminium, one half for tobacco, sugar and petroleum, one third in the case of wheat and three fifths of raw cotton.[18] Without such aid the prospects for the British economy would have been very bleak indeed. A further hefty shift away from dollar imports would have been required which would have led to food rations being cut to well below those ruling during the war; in addition, there would have been a severe check to the supply of consumer goods and a setback to recovery with possibly 1½ million unemployed.[19]

During 1948 the balance of payments improved dramatically

under the influence of a tight domestic policy, rising exports and the flow of Marshall aid. The balance on visible trade, though still in deficit, registered a sharp improvement, as did the invisible surplus, while the dollar deficit was halved (see Table 1). But the situation did not last for long. By 1949 the current-account balance was deteriorating once again, sterling came under severe strain and in September the pound was devalued by 30.5 per cent from $4.03 to $2.80.

The background to the second major external crisis is complex. The annual figures for the balance of payments give no real indication of a severe problem. The current-account balance was only marginally worse than in the previous year, and on later revised

*Table 1: British Balance of Payments, 1946-51 (£mn)*

| | | 1946 | 1947 | 1948 | 1949 | 1950 | 1951 |
|---|---|---|---|---|---|---|---|
| Balance on visible trade | (a) | - 103 | - 361 | - 151 | - 137 | - 51 | - 689 |
| | (b) | - 200 | - 449 | - 218 | - 180 | - 153 | - 789 |
| Balance on invisibles | (a) | - 127 | - 20 | 177 | 136 | 358 | 320 |
| | (b) | - 250 | - 226 | 98 | 110 | 382 | 268 |
| Balance on current account | (a) | - 230 | - 381 | 26 | - 1 | 307 | - 369 |
| | (b) | - 450 | - 675 | - 120 | - 70 | 229 | - 521 |

(a) as estimated in 1980
(b) as estimated at the time

*Dollar Balance, 1946-51 (£mn)*

| | 1946 | 1947 | 1948 | 1949 | 1950 | 1951 |
|---|---|---|---|---|---|---|
| Visible balance | - 290 | - 437 | - 210 | - 247 | - 115 | - 349 |
| Invisible balance | - 11 | - 73 | - 42 | - 49 | 27 | - 87 |
| Current-account balance | - 301 | - 510 | - 252 | - 296 | - 88 | - 436 |

*Sources:* A. Cairncross and B. Eichengreen, *Sterling in Decline* (1983), p. 145 and *United Kingdom Balance of Payments, 1946 to 1956* (No. 2), Cmnd. 122 (1957), Table 2.

estimates it about broke even, while even the dollar deficit was very much less than it had been in 1947 (see Table 1). Moreover, the dollar deficit of the outer sterling area was, if anything, less than it had been in 1948.[20] It is true that Britain was still having great difficulty in balancing her accounts with the dollar area at the ruling exchange rate and only by the use of tight controls on imports from this region was it possible to limit the size of the deficits. However, the partial relaxation of some controls in 1948-49, rising domestic activity and a moderate recession in the United States all helped to weaken the dollar account. Nevertheless, the annual figures by themselves scarcely indicate the size of the problem and one must turn to fluctuations within the year to discover the source of the difficulties.

It was during the second quarter of 1949 that the dollar account, which had been improving for about eighteen months, deteriorated very sharply. In fact the deficit doubled in that quarter, as did the deficit of the sterling area as a whole, and Britain lost over £160 million of gold. This large adverse swing was partly due to a very big increase in British import expenditures in the United States, much of which represented a one-off stockpiling of food and raw materials and would not therefore be expected to continue. Secondly, the downturn in domestic activity in the US curtailed imports from both Britain and the outer sterling area, and full recovery in US import demand did not take place until the first half of 1950. However, despite these exceptional movements and the imposition of import cuts in Britain and the outer sterling area in July, the drain continued. Britain's gold and dollar reserves which stood at £471 million at the end of March and £406 million at the end of June, fell to £372 million by mid-August and £330 million a month later.[21] It is clear that the rapid outflow after the second quarter was based on the speculative belief that the pound would be devalued, a subject known to have been under consideration by the government for some time, since the dollar trade balance was improving in the latter half of 1949.[22] There was a general feeling that the pound was overvalued relative to the dollar and that eventually it would have to be adjusted downwards to rectify the dollar deficit. 'It was the perception ... that devaluation was a necessary ingredient in the restoration of international equilibrium that fuelled speculative pressure; and it was the speculative pressure that in the end compelled devaluation.' It is true that a decision to devalue the pound had been taken before the speculative pressure reached its peak, but the sudden and sharp loss of reserves subsequently 'played a decisive part, and to that extent devaluation was a capitulation to market opinion and another of many demonstrations of the weakness of government in face of an exchange crisis.'[23]

The size of the devaluation was somewhat larger than expected but, as the Chancellor of the Exchequer (Sir Stafford Cripps) explained, it was designed to give British exporters a real competitive edge in the North American market. He also emphasised the fact that it was a final solution rather than a first tentative step. It prompted the realignment of most major currencies including the Canadian dollar, though only a few countries devalued by as much as 30 per cent. The trade-weighted devaluation of sterling therefore eventually emerged at around 9 per cent. The devaluation also led to a readjustment of internal policy. While depreciation was seen as a way of avoiding severe deflation at home, it was nevertheless deemed necessary to tighten the fiscal stance in order to make devaluation effective by releasing resources for export and to prevent undue inflationary consequences. Thus measures to cut home demand were announced in the autumn of 1949, which included cuts in defence spending, education, housing and investment in nationalised industries, together with a series of smaller measures, though it is doubtful whether all of these were effectively implemented.

The following year again saw a significant improvement in the balance of payments for Britain and the sterling area as a whole. The current balance showed a large surplus while the dollar deficit had been reduced to its lowest since the war at -£88 million. In both cases the trade balance improved though the main contribution came from the sharp turnround in invisible items. As the speculative attack subsided the reserves were rapidly restored; three fifths of the loss during the summer had been regained by the end of 1949 and by April of the following year most of the remainder had been recouped. Not all the improvement can be attributed to policy measures however. Dow believes the effect of devaluation was the least influential and attributes more to the sharp rise in US import propensity following the ending of recession in that country.[24] On the other hand, it would seem that devaluation did bring some lasting benefits with relatively small effects on domestic inflation. The competitive edge was maintained, the Korean War notwithstanding, through to the early 1950s at least, when healthy surpluses were being recorded. Moreover, there was a notable shift in exports from sterling to dollar markets and a drop in the share of imports derived from dollar sources of supply, though in part these may reflect a recovery in the sterling area's supply capability from wartime disruption. The shifts persisted into the 1950s.[25] Planned dollar cuts were, however, far less effective than in 1947-48, possibly because many of the discardable dollar imports had already been eliminated.[26]

Unfortunately this marked improvement was rudely interrupted in the final year of Labour's office by the outbreak of the Korean War in June 1950, which sparked off the third major payments crisis of the postwar period. Though in magnitude almost as severe as that of 1947, it differed in some respects from the earlier crisis. For one thing the sharp rise in the deficit, unlike 1947, had not been foreseen, but emerged suddenly as a result of an external shock. Secondly, the main deterioration occurred on the trade account rather than invisibles, except in the case of the dollar account where the invisible turnround was pronounced (see Table 1). The adverse shift in the trade balance was primarily due to a sharp rise in imports, which in value terms were nearly 50 per cent greater than the previous year, composed of a 10 per cent volume increase and a 30 per cent price rise. The import effect was accentuated by the rundown in stocks in the previous year when imports had been restricted, and by the world commodity price boom as a result of US rearmament activity.[27] Exports remained flat for the year as a whole (1951) though initially the US restocking boom had been favourable to British exports which experienced two sharp surges, one late in 1950 and the other in the second quarter of 1951. However, manufacturing exports subsequently faltered so that by the end of 1951 they were slightly lower than a year earlier, which is probably to be explained by the capacity constraints following Britain's own rearmament programme. The British payments position was also exacerbated by large swings in the dollar earnings of the overseas sterling area which fell dramatically in the latter half of 1951, and by the partial convertibility granted to West European sterling balances which had been agreed in negotiations leading to the formation of the European Payments Union in 1950.[28]

Thus Labour's final phase was marred by yet another crisis leading to the re-imposition of some controls on prices and raw materials, together with plans for substantial import cuts which in fact were left to the incoming Conservative government to implement following the October general election.

In view of the strong rise in industrial production and exports in the postwar years (see next section), it is somewhat ironic that Britain should have suffered from periodic crises in her external dealings. But, as with Western Europe, the payments problem was very much a dollar one, a consequence basically of the emergence after the war of one dominant supplier in a world hungry for goods of all kinds. The dollar scarcity was destined to persist well into the 1950s until Europe had fully restored her supply capability. On the other hand, it is also clear that Britain's own payments problem was exacerbated by additional factors. The 1947 crisis was aggravated by the internal dislocations to production, while in the case of 1949 speculative activity in anticipation of a devaluation was very much in evidence. There is also the suspicion, though not fully confirmed, that the government was trying to achieve too much in too short a space of time which led to a level of consumption and investment incompatible with payments equilibrium. Certainly the sharp rise in consumption shortly after the end of the war put too much pressure on the domestic economy. The volatility of the sterling area's fortunes and the liability of the sterling balances also added to Britain's problems. Furthermore, the order of priorities may be open to criticism — for example the high level of government spending abroad, especially in maintaining troops, and the rate of foreign investment, which amounted to some £1,650 million in the six years 1945-51.[29] But when all is said and done, the fact remains, as an ECA study pointed out, that prior to devaluation there was no natural tendency towards equilibrium in the dollar account.[30]

## Industrial Recovery and the Export Boom

Despite the periodic crises, the overall performance of the British economy after 1945 was quite impressive. The main economic indicators in Table 2 show a fairly continuous recovery through to 1951, with levels of activity substantially above those ruling immediately prior to the war. Even in 1947, when the economy was hit by two crises, the coal distribution crisis early in the year and the sterling fiasco in the summer, the economy was more buoyant than the gloomy commentaries of contemporary observers would have us believe. Rates of growth may have been depressed for a time but in fact it was from the middle of that year that the great postwar upsurge in production really gained momentum following the transition to peacetime production. This was achieved remarkably quickly and smoothly for by the end of 1946 some 3¼ million engaged in war industries had been switched to civilian production, while a large part of the armed forces had been demobilised. The transfer was aided by the high demand for labour and by the careful planning which had gone into demobilisation during the war.

In every respect the experience was so much better than that after the first world war. This time there was no sharp boom and slump after which production and exports languished below their 1913 levels and unemployment soared to over two million. Nor was there a violent price boom or frothy speculative activity as economic agents bid for commodities and assets following the hasty abandonment of economic controls in 1919-20. Instead there was a strong and steady expansion in output and exports, while controls kept inflation and consumption in check and employment remained high. Europe too experienced a strong recovery at this time, again in marked contrast to the disastrous inflationary spiral of the early 1920s. In fact Europe's progress after 1948 was better than the British and indicated a foretaste of things to come. Apart from the United States, Australia and Canada, Britain was the only major country to emerge from the war with a level of output higher than in 1938. All other countries recorded falls between 1938-45, ranging from over 50 per cent in the case of Austria and Germany to less than 5 per cent for Norway and Denmark.[31] Subsequently, rates of recovery were so strong that by 1951 Denmark, Finland, France, the Netherlands, Norway and Sweden had surpassed their 1938 levels of GDP by a larger margin than Britain; Italy and Belgium

*Table 2: Main Economic Indicators, 1938-51 (1938 = 100)*

| | 1946 | 1947 | 1948 | 1949 | 1950 | 1951 | percentage increase 1946-51 |
|---|---|---|---|---|---|---|---|
| Gross domestic product | 111.5 | 109.9 | 113.4 | 117.6 | 121.4 | 125.0 | 12.1 |
| Industrial production | 102.5 | 107.9 | 117.2 | 124.0 | 131.1 | 135.4 | 32.1 |
| Manufacturing | 104.6 | 110.7 | 120.7 | 128.4 | 137.2 | 143.1 | 36.8 |
| Volume of exports | 100.0 | 109.6 | 138.5 | 153.9 | 175.0 | 173.1 | 73.1 |
| Volume of imports | 67.4 | 76.1 | 78.3 | 84.8 | 84.8 | 94.6 | 40.4 |
| Consumers' expenditure: | | | | | | | |
| Total | 99.8 | 102.9 | 103.9 | 105.8 | 108.8 | 107.3 | 7.5 |
| *Per capita* | 96.3 | 98.2 | 98.2 | 100.0 | 101.9 | 101.9 | 5.8 |

*Sources:* C. H. Feinstein, *National Income, Expenditure and Output of the United Kingdom, 1855-1965* (1972), T19, T22, T42, Cambridge University Press; London and Cambridge Economic Service, *The British Economy: Key Statistics, 1900-1966*, p. 14.

were not very far behind while Germany and Austria were well on their way to recovery. Only Japan's output remained below the prewar level.

Thus despite a strong performance there were signs later on in the period that Britain was being overtaken by her main competitors, a situation which began to be reflected in the export trade. Exports bounded along at a prodigious rate to reach a level in 1950 75 per cent above prewar, the target set as the minimum required to achieve equilibrium but which no one at the time anticipated could be met. Exports in fact rose very much faster than GDP or industrial production, while import volumes were severely contained. From the data in Table 2 it can be seen that the main casualty was consumption which, after a sharp rebound in 1945-46, only rose very slowly through to 1951. In fact *per capita*, consumers' expenditure was only marginally higher than before the war.

A more detailed breakdown of the output data by sector of activity is shown in Table 3, which gives a broad idea of where the main growth was coming from. Generally speaking, the dynamic sectors were those which had done well in the war. Thus above average rates of expansion were recorded in chemicals, metal manufacturing, the engineering trades (including electrical), other manufacturing, gas, electricity and water and transport and communication. Some of these were relatively new sectors with good potential growth prospects where innovation, investment and productivity growth were fastest. These sectors did much to account for the rising trend of productivity and capital per worker in the economy as a whole. Productivity in industry (excluding building) rose sharply after 1947 at nearly twice the prewar rate, and by 1950 it was about 20 per cent higher than in 1938.[32] By 1948 gross investment had already surpassed the prewar level and the net capital stock had probably recovered to its former level. By contrast, some of the older industrial sectors, including mining, building and contracting and textiles, while showing a strong recovery between 1946-51, still languished at or below their peacetime levels of activity. The growth of most of the service trades was constrained, as had been the case in wartime apart from the government sector.

Among the growth sectors there were some striking success stories. The most spectacular was that of motor-car manufacturing which the government called upon to streamline its operations and improve its efficiency in order to boost exports. In 1945 there were about 30 firms making some 100 different models, but by 1949 the number of models had been cut to 40 produced by 23 firms. The number of vehicles turned out rose rapidly after 1946 and by 1950

*Table 3: Indices of Output by Sector 1938-1951 (1938 = 100)*

| | 1946 | 1947 | 1948 | 1949 | 1950 | 1951 | % increase 1946-51 |
|---|---|---|---|---|---|---|---|
| Agriculture, forestry and fishing | 110.7 | 106.2 | 114.2 | 122.8 | 125.7 | 128.5 | 16.1 |
| Mining and quarrying | 77.8 | 80.1 | 84.9 | 87.7 | 88.6 | 91.6 | 17.7 |
| Chemicals and allied trades | 150.2 | 152.6 | 165.2 | 170.8 | 193.8 | 203.5 | 35.5 |
| Metal manufacturing | 123.8 | 131.8 | 147.2 | 147.9 | 155.4 | 164.9 | 33.2 |
| Engineering and allied trades | 119.8 | 127.0 | 137.9 | 147.8 | 157.6 | 167.5 | 39.8 |
| Textiles, leather and clothing | 73.8 | 79.3 | 89.2 | 96.3 | 103.0 | 101.1 | 37.0 |
| Food, drink and tobacco | 110.4 | 112.4 | 115.1 | 119.6 | 118.6 | 122.6 | 11.1 |
| Other manufacturing | 90.0 | 98.0 | 109.0 | 119.1 | 132.2 | 140.1 | 55.7 |
| Building and contracting | 76.8 | 80.1 | 87.3 | 91.3 | 91.4 | 87.9 | 14.5 |
| Gas, electricity and water | 134.0 | 138.9 | 147.7 | 157.9 | 172.2 | 182.9 | 36.5 |
| Transport and communications | 114.1 | 119.2 | 128.2 | 131.3 | 134.2 | 140.3 | 23.0 |
| Distributive trades | 81.5 | 88.4 | 91.9 | 97.9 | 102.2 | 100.4 | 23.2 |
| Insurance, banking and finance | 97.8 | 103.3 | 104.2 | 106.1 | 111.6 | 112.6 | 15.1 |
| Professional and scientific services | 102.2 | 106.0 | 111.4 | 116.0 | 122.1 | 125.2 | 22.5 |
| Miscellaneous services | 84.5 | 85.8 | 87.1 | 83.1 | 81.8 | 81.8 | -3.2 |
| Public administration and defence | 239.7 | 175.5 | 153.1 | 148.6 | 144.1 | 151.6 | -39.8 |

*Source:* C. H. Feinstein, *National Income, Expenditure and Output of the United Kingdom, 1855-1965* (1972), T25, Cambridge University Press.

was nearly three quarters above the prewar level when it accounted for about one half the European output. Since American cars were not readily available because of the dollar shortage and European output was slow to recover, British producers found themselves in a seller's market, and domestic sales were restricted to take advantage of the strong export market. Export sales responded well, with no less than 69 per cent of output going abroad in 1950. Even so there is some evidence that the industry became unduly complacent in this buoyant market, too much attention being placed on design of product for the Dominion and domestic markets rather than for the North American. Moreover, despite substantial productivity gains, British makers still fell well behind their American counterparts in efficiency.

Other newer industries to register substantial progress included aircraft manufacturing, chemicals and electrical engineering. The aircraft industry had come of age in the war and it maintained its international lead in the postwar years with assistance from the government and a strong home market. Britain led in the development of gas turbines and jet engines and in their use in civil air transport. British aircraft manufacturers led the world in the number of new types developed in 1950. Similarly, the chemical and allied trades benefited from the temporary eclipse of Germany as a competitor in world markets and took advantage of the growing demand for new products in plastics, pharmaceuticals and synthetic fibres. The output of this branch rose by a third or so to a level twice that of prewar.

One of the most crucial intermediate products in the reconstruction process was iron and steel since cheap and abundant steel formed an input in the case of so many industries. In recognition of this the British Iron and Steel Federation submitted a report to the government in 1946 calling for a substantial expansion and improvements in capacity, including a large-scale programme of investment at strategic locations. Altogether capacity was planned to expand by 40 per cent but many old and inefficient plants were to be retained in the interim because of the anticipated heavy demand for steel. In the event, the adaptation of the industry can be criticised for its patched-up nature and the sub-optimal location of new plants, but in the short term at least it delivered the goods. Steel production surpassed the government's target by 1950 when it was 60 per cent above prewar, and steel rationing was terminated in May of that year except for steel sheets.[33]

Some of the older industries and those producing primarily for the civilian market made much less headway. As in wartime, the coal industry, the first to be nationalised, again performed badly. Though output recovered slowly from the low point reached in 1945 it remained below the 1938 level right through to 1951. The industry was therefore unable to keep pace with demand, the most striking illustration of this being in the winter of 1946-47 when the economy almost ground to a halt because of the fuel shortage. Basically it was the same old story: low productivity, lack of investment, deteriorating geological conditions and a recalcitrant workforce. It was also found increasingly difficult to attract workers to the industry at a time when there were more enticing jobs elsewhere in the economy.

Textiles also found difficulty in rebuilding their workforce following the sharp cutback in wartime, partly because of the rather unattractive rates of pay. As a result output of the industry barely managed to recover to the peacetime level by the end of the period, by which time export markets were already weakening as competition increased from Japan and third-world suppliers. Rationing of clothes remained in force for some years because of supply shortages and the needs of the export trade. But generally the textile industries missed their opportunity in these years; they still suffered from the legacy of interwar depression — outdated equipment and low productivity — with the result that they never had a chance of recovering their former glory. The one bright spot was the rapid expansion in man-made fibres — rayon, nylon and terylene — though such products are perhaps more appropriately classed as offshoots of the chemical industry rather than textiles proper.

Building and contracting also found great difficulty in attracting sufficient workers in the tight labour market of this period. Output therefore only increased slowly and was still well below the prewar figure by 1951. Tight restrictions on inessential building work and a severe shortage of imported materials, notably timber, restricted the rate of expansion, though Britain managed to devote a higher share of her investment to housing than virtually any other European country.

Finally, one might make reference to agriculture again, which continued to build upon its wartime success. Still considered a vital industry because of world food shortages, especially grain since bread had to be rationed in 1946, and the need to conserve scarce foreign currency, the preference accorded to the industry during wartime was made permanent by the Agriculture Act of 1947. This provided for annual price reviews and guaranteed markets for the major products. As a result farmers continued to do well, and output expanded to some 30 per cent above prewar by the early 1950s, a much better record than that of any other European country.[34]

Even more impressive than the output record of British industry was the spectacular growth in exports. From the low point recorded in 1944, when they were but 30 per cent of the 1938 level, exports had regained their prewar level by 1946 and by the turn of the decade had surpassed former peacetime levels by 75 per cent. What is more, the composition of exports was changing for the better. The relative importance of the old staple exports was declining in favour of the newer and rapidly expanding sectors of activity in world trade, that is chemicals, electrical machinery, road vehicles, petroleum products, pharmaceuticals, new fibres and the like. By the early 1950s some two thirds of Britain's exports consisted of those products which were expanding in world trade.[35] This success no doubt owed something to the government's export drive backed up by the controlled allocation of raw materials and the restriction on domestic consumption.

However, despite this impressive achievement there were some disquieting features about Britain's trading record which boded ill for the future. The upsurge in exports in the early years after the war is not altogether surprising given the low level of European output, the severe control on dollar imports and the limited competition from third-world producers. British exporters had a seller's market at their feet and should therefore have done well. But by the end of 1948 and the first half of 1949 there were signs that export volumes were flattening out as European countries recovered, and in the years 1948-51 European exports were growing faster than the British. British export volumes received a renewed fillip late in 1949 and 1950 with the ending of the US recession and the devaluation of sterling, but it proved to be a temporary affair after which they stagnated until the mid-1950s.

Thus it appears that by the end of the decade or thereabouts British exports had exhausted the once-and-for-all bonus stemming from the reduced supply capability of the main competitors in the international market. Once their capacity to produce and export was restored Britain no longer had the advantage of a seller's market and her former inability to compete tended to reassert itself despite the observed shift in the export product mix and the bonus of devaluation. Moreover, though the external account was in equilibrium in 1950, this was only made possible by continued tight control on imports, especially those originating from the dollar area. The volume of imports, though rising slowly in the postwar years, remained below the prewar level throughout despite the higher level of domestic activity, and there was a marked shift away from dollar sources of supply after 1947. In fact import volumes remained below the prewar level until the middle of the 1950s since the Korean War crisis necessitated further sharp cutbacks in volume. In other words, far from permitting a larger volume of imports, much of the increase in exports had been absorbed in closing the gap left by the loss of income on invisibles and the mounting debt costs, together with making good the deterioration in the terms of trade. Consequently, if Britain were to maintain solvency and ensure a satisfactory level of imports, it was estimated that the volume of exports would have to be at least double that of 1938, and that the UK's share of world trade would need to be considerably higher than before the war.[36] The prospects of achieving this revised target were far from promising in the more competitive climate of the early 1950s.

Even more disturbing from a long-term point of view was the market structure of Britain's exports. While the commodity composition of Britain's exports had altered for the better, the same could not be said of their geographical distribution. Britain continued to concentrate on the downstream markets of the sterling area and the third world to the neglect of the rich and expanding markets of North America and later Western Europe. In fact by 1949 the proportion of exports going to the latter areas was less than before the war, while over half went to the sterling area as against 45 per cent previously (see Table 4). There were several disadvantages in concentrating on extra-European markets: they were more volatile because of their dependence on primary incomes, there was a greater threat of import substitution, and they were less demanding in terms of product requirements and therefore less likely to exert a favourable influence on technological progress and productivity growth in the supplying country. In short, they were more likely to have a debilitating influence on Britain's ability to compete elsewhere in the future since they would tend to discourage a move upstream towards more sophisticated, high unit value products. At the same time one should recognise the fact that there was some justification for concentrating export activity on these areas in the short term. There was a high level of unsatisfied demand in the early years after the war, an already well-established and comfortable pattern of trade and payments relationships, and of course the prospect of earning dollars which could not be obtained in Europe. The overhang of the large sterling balances also tended to strengthen the trend, as did the source of many food and raw material imports. On balance, therefore, it was found considerably easier to push exports to these markets than to North America or Western Europe.

In the immediate term the most crucial market from Britain's point of view was the North American, since here the scope for earning dollars was potentially greatest. But the market proved a difficult one to exploit because of the self-sufficiency of the US and the latter's trade policy, but even so the British export pattern scarcely does justice to the importance of this market. Through to 1949 the share of British exports going to North America was well down on prewar at under 10 per cent (Table 4), and the main commodities exported hardly consisted of the most promising items — pottery, whisky, cotton yarn and cloth, linoleum, cotton waste and rags, wool yarn, woollen and worsted manufactures, knitted goods, rayon goods and books. The pattern of trade in fact reflected a bygone era and one may note the absence of sophisticated new products, capital equipment and high unit value exports. As Milward notes, not a single one of these leading commodities featured among the 20 most valuable categories of US imports.[37] It is true that the intensification of the export drive after 1947 resulted in greater attention being given to the dollar area and in 1949 a Dollar Export Board was established for this purpose. The result was that dollar export volumes increased quite significantly, with a sharp jump in 1948 and a further rise of 50 per cent between 1949-51. But progress in later years was partly fortuitous, owing a great deal to devaluation, the revival of US import demand following recovery from the recession, and later the demand generated by the Korean War effects. There is no indication that there was a strong and

*Table 4: Geographical Distribution of Britain's Exports, 1938-51 (percentage shares)*

| | 1938 | 1946 | 1947 | 1948 | 1949 | 1950 | 1951 |
|---|---|---|---|---|---|---|---|
| North American | 10.0 | 7.5 | 8.1 | 8.7 | 7.6 | 11.0 | 10.6 |
| Other American account countries | 1.6 | 1.4 | 1.8 | 2.1 | 1.7 | 2.0 | 2.0 |
| Rest of sterling area | 45.0 | 45.1 | 48.7 | 49.7 | 52.0 | 48.0 | 50.7 |
| OEEC countries | 25.0 | 28.3 | 23.4 | 23.4 | 22.3 | 25.0 | 24.2 |
| Rest of world | 19.4 | 17.7 | 18.1 | 16.1 | 16.4 | 14.0 | 12.5 |

*Source:* E. Zupnick, *Britain's Postwar Dollar Problem* (1957), p.102, © 1957, Columbia University Press. By permission.

lasting real improvement in dollar export capability and little evidence of a dramatic shift in the export product structure.

The missed opportunity in Western Europe was even more ironic given the fact that until the middle of 1949 at least Britain had it within her grasp to participate in the reshaping of Europe and its recovery thus consolidating her position for the future. Unfortunately the Labour Government eventually turned its back on Europe in favour of the time-honoured Imperial connections. Consequently Britain failed to participate in the trade boom within Western Europe between 1948-51, which helped to insulate the latter from the main effects of the US recession of 1949 and the payments crisis which hit Britain.

In sum therefore, while Britain's export performance was good in the reconstruction period it was not good enough given the loss of overseas income, the debt costs and the import needs of a growing and unrestricted economy. The product mix of Britain's export trade did move upstream overall, towards more advanced products, though much less so on dollar account, but there continued to be persistent doubts about her underlying competitiveness in world markets.[38] Devaluation eased those doubts but did not extinguish them. The shift in product structure was not however matched by a commensurate switch in the market areas. International payments difficulties and the attachment to the connections of a bygone age left Britain with a market structure distinctly out of kilter with later needs. 'The pattern of Britain's external economic relationships was set throughout the prosperous 1950s in the unsatisfactory mould to which exigency had shaped it before 1949.'[39]

## Nationalisation and the Welfare State

Public ownership of key industries and the reform of the social services were the two key items in Labour's structural policy programme. Whatever the cost it was determined to push these through partly for political purposes. Whether either were crucial to the reconstruction effort is to be doubted. In fact, given the pressure on resources and on ministers' time and energy it could be argued that they were something of a luxury, at least in the early years. And bearing in mind a later analogy, nationalisation probably had as much relevance to the tasks of reconstruction as Mrs. Thatcher's experiment in privatisation was to have to the recovery effort in the early 1980s. Nevertheless, the policy promises were kept and large-scale public ownership and a comprehensive welfare state became a reality.

Nationalisation of basic sectors was seen to be the cornerstone of Labour's drive to socialisation as a means of eliminating the stagnation and unemployment of the past while at the same time dealing with those ailing industries which had 'failed' the nation. The programme rolled forth as fast as the Bills could be drafted in the years 1946-49 and it proved to be the most time-consuming and controversial part of Labour's legislative feast. In 1946 no less than four major sectors were dealt with: civil aviation, Cable and Wireless, the coal industry and the Bank of England; electricity and railways, including long-distance road transport, followed in 1947; gas in 1948 and iron and steel in the following year. By the time the steel debates were out of the way the Labour Government's enthusiasm for public ownership was visibly waning.[40] In the election manifesto of 1950 the issue featured much less prominently, though it did contain a motley 'shopping list' which was conveniently forgotten when Labour resumed office. The election manifesto of the following year spoke vaguely about the merits of public ownership in terms of the national interest but did not specify anything in particular.

Nationalised undertakings or semi-public corporations were of course nothing new. There was a whole series of such examples prior to 1940, including the Post Office of long standing, and the series of measures in the interwar years setting up the BBC, the Central Electricity Board, The London Passenger Transport Board and British Overseas Airways Corporation. What was novel about the postwar programme was its sheer size and the doctrinal considerations determining it. Altogether some 20 per cent of industry was transferred to state ownership in the period, embracing a total workforce of around 2½ million (including the Bank of England), the largest of which were inland transport and coal which accounted for 1.7 million of the total. However, because of the size of the programme and the speed with which it was implemented, insufficient attention was given to the organisational structure and the operational procedures of the new undertakings. Space forbids a detailed examination of each entity and we shall therefore concentrate on some of the general issues and problems which are common throughout.

The government had no clear blueprint for the organisational format of its intended acquisitions and so the structure of each undertaking varied considerably depending upon individual circumstances. In the case of coal, a monolithic structure was imposed, with the National Coal Board (NCB) responsible for the numerous properties acquired and operating through a line-and-staff form of organisation. Regional and divisional boards were subsequently set up by the NCB for its own convenience, and these derived their authority directly from the delegated power of the National Board. By contrast, statutory area boards were set up for both electricity and gas. The twelve area boards in electricity were responsible for the distribution of power, subject to the authority and financial control of the British Electricity Authority (BEA), which directly controlled the generation of electricity and the national grid system. The gas industry had a similar type of organisation, but in this case the twelve area boards owned and operated all gas works and distributive systems and were directly responsible for balancing their own accounts. There was very little coordination among them and the National Gas Council had much more limited powers than BEA; it was to function largely as a consultative and advisory body on behalf of the area boards with power to intervene only if a board failed to meet its financial obligations. But by far the most complex structure was that for transport, partly because of the wide range of activities which were taken over. The British Transport Commission (BTC) had responsibility for the general principles and planning of transport, but provision was made for the establishment of five executives, that is one each for the railways, docks and inland waterways, road transport (later split into two), London transport, and hotels, each of which managed the specific properties in question and was directly accountable to the BTC. In turn, the executives could organise their work on regional lines, subject to the Commission's consent, and in the case of the railways six regional divisions were eventually established by the executive. As for aviation, the Civil Aviation Act of 1946 reversed the fusion policy of 1939-40 by slicing BOAC into three groups each with a clearly identified sphere of operation. BOAC was to concentrate on main haul routes including those of the Commonwealth, the Far East and North Atlantic; internal and European air routes were entrusted to British European Airways (BEA); while British South American Airways Corporation (BSAAC) had the task of developing the South American market.[41] The three state corporations were autonomous and had exclusive rights to operate scheduled services in their designated areas. Finally, the Iron and Steel Corporation dealt directly with the individual companies or groups acquired, though the Iron and Steel Federation, representing both public and private interests, continued to exercise considerable authority in the industry which led to complications in the pattern of organisation and control.

Apart from specifying the organisational structure, the nationalisation Acts laid down various duties and conditions which the undertakings had to observe. Each industry was enjoined to provide an efficient, reliable and economical service: thus in the case of transport, for example, the BTC was charged with the duty of providing 'an efficient, adequate, economical and properly integrated system of public inland transport and port facilities within Great Britain for passengers and goods with due regard to the safety of operation'. On financial matters the state industries were expected to be self-supporting, making charges to customers sufficient to cover costs of production, administration and debt servicing, 'taking one year with another', a suitably vague phrase which left room for much misinterpretation. Thus the Coal Nationalisation Act of 1946 laid down that 'the policy of the Board shall be directed to securing ... that the revenues of the Board shall not be less than sufficient for meeting all their outgoings properly chargeable to revenue account ... on an average of good and bad years'. This general principle of breaking even was followed in subsequent Acts except that in the case of gas the obligation applied to each board rather than the collective undertaking.

Unfortunately the ability to attain the financial targets was circumscribed by the limits set to charging powers. In most cases provision was made for ministerial intervention with respect to the pricing of goods and services, and in the case of the BTC all revisions of charges had to be vetted by the Transport Tribunal, the successor to the old Railway Rates Tribunal. In practice ministerial direction to contain price increases was quite frequent with the result that, at a time of rising costs, the undertakings found it difficult to balance their accounts. Gas, electricity and transport charges in particular were kept artificially low in the national interest and to ease the acceptance by the TUC of the proposed wage freeze in 1948. As a consequence, the financial results of the relevant boards suffered, while heavy interest charges on outstanding debt proved an additional burden.

While limiting the freedom in charging matters, the Acts gave little clue as to how charges should be formulated, whether marginal or average cost pricing should be used. Because of their size and complexity and the difficulties in some cases in determining marginal costs for specific services, the practice was to use average cost pricing, which inevitably provided scope for extensive cross-subsidisation especially in transport. This militated against efficiency and the rationalisation of services and capacity and meant that marginal capacity was kept in being when it had ceased to produce a viable return. There was little incentive, for example, to eliminate uneconomic branch railway lines so long as these could be subsidised by the more profitable parts of the route network, or even by other transport services.

Each industry was expected to plan for the future, and to so arrange their investment plans to meet anticipated demand with up-to-date equipment. In the short term however they were so busy dealing with immediate problems, including setting up their organisational structure, that there was little time for long-term planning, so that by the turn of the decade only the coal industry had published a long-term programme for investment. Moreover, though many of the nationalised industries were in urgent need of new capacity and renovation (e.g. coal, railways), long-term plans had to be shelved or delayed because of the periodic cuts in projected investment in the crisis years of 1947, 1949 and 1951. Thus the reality of the national plan for 1948-53, in which investment in fuel and power and transport and communication was to absorb one third of total gross investment, did not come up to expectations. The Gas Council was allowed very small funds for new installations, the British Electricity Authority, much of whose capacity was 'old, inadequate and costly to operate',[42] was not able to increase the capacity of the national grid as planned until 1953, while the railways were literally starved of investment resources until the 1950s. The coal industry fared somewhat better primarily because of the fear of another fuel crisis like that of the winter of 1946-47.

At the time of acute pressure on resources and periodic external crises, and when the newly acquired industries were struggling to get their act together, it is perhaps not surprising that attention should be focused on short-term issues with a promise of high return. The severe upheaval and disruption at managerial level caused by the massive merger operations were alone sufficient to delay the preparation of long-term plans. Even less thought was given to questions of the integration of services between competing products as in the case of transport and energy.

Short-term problems were not the only reason for the lack of progress in this respect. There is evidence that the formal institutional structure created by the legislation and the obligations imposed on the undertakings were not conducive to promoting the objectives specified. Transport provides a good illustration in this context. It is difficult to see how the BTC could have been expected to provide 'an efficient, adequate, economical and properly integrated' system of inland transport in any full sense of the term given the conditions laid down in the Transport Act of 1947. For one thing the machinery of organisation did not lend itself readily to the integration and coordination of services because, until the early 1950s at least, the structure was highly functional and overcentralised with final power in the hands of the Minister of Transport who could overrule the decisions of both the Commission and its subordinate organs. This meant that schemes for the coordination of services at the lower level had to pass right up the chain of command and on to the Minister before a decision was finally made, a process which was bound to cause both delay and frustration. Moreover, at the executive and regional levels integration was unlikely to be encouraged since each executive tended to organise its own regional structure differently. Eventually the railways had six regional divisions, the Road Executive between 10-12 and the Waterways Executive five.

Perhaps even more serious were the incompatible obligations imposed on the Commission. Presumably to achieve an economical and integrated system of transport some method was required to ensure that each traffic was allocated to the mode of transport that could move it as efficiently and cheaply as possible. But the Commission's room for manoeuvre in this respect was severely limited since transport users were completely free to choose between alternative modes; the BTC did not have a complete monopoly of all inland transport services, but it was expected to promote regional schemes for coordination of passenger transport, whether by road or rail. The only real solution to this problem was therefore to price all services on the basis of costs of operation in the hope that each traffic would select that service least costly in real terms. But since the Commission was also enjoined to provide adequate services and to practice no undue discrimination between customers, this presupposed a pricing policy which departed from the above principle, together with the retention of uneconomic services and cross-subsidisation, a situation the more likely to occur given the fact that the individual executives were not separately accountable. The 1947 Act in fact did little if anything to alter the unscientific pricing policy of the past in the case of the railways, so that the way was left open for cross-subsidisation both within and between transport services. No attempt was made to devise a charging policy related in any rational way to costs and 'the failure to produce a new charges scheme which could meet this need was probably the most serious and fundamental deficiency of the nationalised transport sector.'[43]

Finally, the Act gave scant recognition to the needs of modernisation and technical improvements in transport. Though the Commission had a duty to provide an efficient system of transport, it was not made clear how this was to be achieved or what resources would be available for doing so. In short, the Transport Act was unlikely to produce a viable and efficient system of transport, and it was most defective with regard to the railways. They required modernisation and a pricing system which properly reflected costs so that the system could be rationalised and made competitive with road transport. But such changes were not likely to be brought about by legislation which made little reference to costs, limited the railways' freedom of charging, continued outdated statutory obligations and which left the way open for cross-subsidisation and the retention of uneconomic services on a grand scale.

Whatever justification there may have been for public ownership of specific industries, Labour's legislative programme was probably too ambitious. No doubt it was an important element in boosting the Party's morale, but the programme was too large and too ill-conceived, while it bore little direct relevance to the needs of the time. That many of the acquired properties went on to greater success does not alter the fact that the record through to 1951 left much to be desired. They did not prove the much vaunted panacea once hoped even in the field of labour relations, and the Labour Party itself, admittedly under a barrage of criticism from the press and the Tory opposition, began to lose faith in public ownership in later years of office.

By contrast, Labour's enthusiasm for the welfare state showed no signs of waning. The creation of a comprehensive welfare service was perhaps its greatest and most enduring achievement and one with which it is most readily identified even today. 'By 1951, a plausible updated version of a land fit for heroes had been built on the scarred foundations of an ancient and war-ravaged community.'[44] It is easy, however, to exaggerate the novel and revolutionary features of the welfare programme of the Attlee Administration. Since the days of Lloyd George in the Liberal Administration prior to the first world war there had been a steady growth in the provision of welfare services, while during the second world war the greater availability of social services and the concern for the health and well-being of the nation were an integral part of the war effort. What was more, the main basis of the new welfare state was formulated in the reconstruction plans of the Coalition Government which foreshadowed many of the principal changes in the later 1940s.[45]

What Labour did was basically to codify and make comprehensive the provision of welfare services so that social security could be seen as the birthright of every citizen whatever his or her age. The measures which formed the basis of the new system were passed during the first year of office: The National Insurance Act, the National Health Service Act and the National Insurance (Industrial Injuries) Act. All of these came into operation on the appointed day, 5 July 1948, the year in which the final major piece of legislation, the National Assistance Act, reached the statute book.

The national insurance scheme followed the Beveridge proposals fairly closely. One consolidated scheme provided insurance cover against sickness, unemployment, old age, disablement and industrial injuries, together with benefits for maternity, death, widows and orphans and children (family allowances).[46] The payment of a single insurance stamp gave entitlement to universal coverage provided by the national minimum benefit rates set for each service. The only problem was that the benefit rates did not keep pace with inflation, thus violating the basic tenet of adequacy preached by Beveridge. As a consequence, the national assistance scheme, introduced in 1948 to cover only a small minority who fell outside the scope of the main provisions, was called upon to an increasing degree to supplement benefit rates which were too low. By the end of 1950 over 1.3 million persons were receiving additional benefit under the 1948 Act, one half of whom were old-age pensioners.

The inauguration of the national health service was a more complicated task since it met with considerable opposition on the part of the medical profession, while it involved the consolidation of a wide variety of institutional provision of hospital services, many of which were short of staff, equipment and lacking in adequate medical facilities. Moreover, as soon became apparent, there was a considerable latent demand for medical services once they were provided free of charge, which not only pushed up costs more rapidly than originally anticipated, but also led to excessively long waiting lists for treatment. Private medical practice was not however abolished and in view of the poor facilities and service under the state system in the early years a small but increasing minority of people chose to pay for superior treatment offered by the private sector.

The third main area of the Labour Government's social programme was the provision of more and better housing. The collapse of the building industry during the war and the increasing rate of family formation resulted in an acute shortage of housing which, together with war destruction, may be estimated anywhere in the region of 3-4 million units depending on definitions. In addition, there were several million war-damaged properties that required repairing to a greater or lesser degree. Even had it been possible to attain the prewar rate of housebuilding, some 350,000 per annum, it would have taken a decade or more to clear off the backlog.

The Labour Government was supremely conscious of the need to boost the rate of housebuilding, not least because of the recollection of the housing fiasco following the first world war. Unfortunately, the collapsed state of the building industry and the severe shortage of men and materials, especially timber, meant that only a trickle of new houses came on to the market in 1946. Towards the end of that year however, a new sense of urgency was injected into the housing programme and a substantial drive was made to get more built, including temporary prefabricated units. By the following year, therefore, nearly 140,000 permanent houses were completed as against just over 55,000 in the previous year, and in the next four years an average of over 200,000 a year were built, making a total of just over one million permanent homes in the period 1945-51.[47] In addition, some 150,000 'prefabs', designed as temporary dwellings but often lasting into the 1970s, were erected, together with a rather larger number of conversions and repairs of damaged properties. Most of the new dwellings were built by local authorities and let at artificially low rents. In all, some 22 per cent of total gross domestic fixed capital formation went into housing in the years 1947-51,[48] a proportion often regarded as excessive in view of the competing claims on resources. In fact, Britain managed to devote a greater proportion of fixed investment to residential accommodation than most other European countries in the reconstruction period, despite the fact that she suffered less damage to her housing stock with the exception of neutral countries.[49] However, it is not easy to pass such judgement when one recalls the size of the housing shortage and the fact that many people were still living in what could only be regarded as sub-standard accommodation.

In educational matters Labour proved to be much more conservative. It accepted more or less without question the provisions of the Butler Education Act of 1944 which provided for a comprehensive and free system of state secondary education for 11-15 year olds, with its commitment to a tripartite division into grammar, modern and technical schools and the eleven-plus examination, all of which were seen to be socially and educationally divisive by later educational reformers. It resisted any pressure to alter or modify the system, while in other areas of education the Party proved equally impervious to reform. The public schools and the universities were left undisturbed, though there was some reduction in the number of direct-grant schools. Apart from the improvement in facilities for elementary education and the emergency postwar teacher-training scheme, education was one area 'where the Labour government failed to provide any new ideas or inspiration.'[50]

In addition to the major areas of social welfare, there were a number of minor reforms dealing with minority groups and interests, for example the blind, mentally handicapped, personal rights, and recreational facilities. In total these were small compared with the universal schemes and it is the latter on which one needs to concentrate when assessing the impact and significance of Labour's welfare programme.

There can be no doubt that the majority of the population benefited from Labour's welfare provisions. The extensive use made of the national health service in its first year of operation is evidence enough that the facilities were universally welcomed. Spending on social services in real terms was higher in this period than it had been before the war, much of the increase being accounted for by the health service. By the early 1950s public spending *per capita* annually on social services was about £11 (at 1900 prices), as against £6.7 in 1938, an increase of 64 per cent, though to put the record into longer-term perspective one should note that during the interwar period it had doubled from £3.3 to £6.7 *per capita* (1920-38).[51] On the other hand, a good part of the increase in the earlier period was on account of unemployment, whereas this was not the case in the full-employment conditions of the postwar years. By 1950-51 the percentage of government spending absorbed by social services was between 43-46 per cent, a figure not dissimilar to that of the 1930s, but in terms of GNP it was much higher, 18 per cent compared with about 11 per cent prewar.[52]

In other words, the Labour welfare programme did produce a significant quantum addition to the nation's consumption of social services. However, this cannot be seen wholly as a transfer of income or payment in kind from the richer to the poorer sections of the community. Many middle-class families benefited from the services, in some cases more so than the poor. The most obvious example is education especially where this involved a direct switch from public fee-paying schools to the state system. Moreover, many working-class families probably paid more in taxes and contributions than they derived in welfare benefits. Had the system of benefits and payments been more discriminatory in terms of income levels, then the scale of the transfer would have been of greater significance. As it was, the main transfer effect was not so much from rich to poor, but from healthy single persons to large families, the aged and the sick.[53]

Nor should one get too excited by the scale of provision laid down. The net benefit rates set hardly constituted a King's ransom — 26s. for sickness, the same for unemployment, 25s. for a single pensioner and 42s. for married couples. These were little more than subsistence levels and their real value was eroded over time by inflation, with the result that an ever-increasing number of applications were being made for national assistance. Thus the welfare measures by no means abolished poverty, and even in later years, when improvements were made to the basic system and cash benefits were adjusted, there still remained a significant residual element in society. Perhaps no system of welfare can cope with the ever rising threshold concept of poverty, and if Labour's programme did not finally abolish poverty, it provided a solid foundation on which to build in the future.

## Wages, Prices and Consumption

In the difficult economic climate of the postwar period the need to moderate the growth in wages and consumption was not in dispute. The top priorities were investment and exports and, as in wartime, resources had to be diverted away from consumption, while wage and cost pressures had to be contained if Britain was to remain competitive in world markets.

At first glance the postwar record looks to have been remarkably successful. The data in Table 5 show that throughout the years 1945-51 earnings (both nominal and real) and prices rose less on average than they did through the war period, though one should note that in the latter half of hostilities the pattern was far more stable than at the beginning of the war. The opposite is the case however with consumption, which had fallen continuously until 1944. Thus the sharp rise in the last year of war and the first year of peace can be seen as a reaction to the very depressed state previously. Thereafter the growth in consumption was relatively modest, and by the end of the period it was not very much above the prewar level.

Some slippage of consumption was only to be expected in the immediate postwar period as consumers sought to work off their accumulated liquid balances. But lax monetary and fiscal policies in 1945-46 certainly contributed to the upsurge in spending. Dalton's first two budgets reduced taxes and probably increased consumer spending power by an amount corresponding to 6 per cent of national product leading to an increase in consumer demand of about 4 per cent of total product.[54] Since extensive price controls and subsidies held inflation at bay the excess purchasing power

*Table 5: Wages, Prices and Consumption, 1938-39 – 1951-52 (annual percentage changes)*

| | Average weekly wage rates | Average weekly wage earnings | Retail prices | Real consumers expend-iture | Real consumers expend-iture *per capita* |
|---|---|---|---|---|---|
| 1938-39 | 1.1 | — | 3.3 | -2.8 | -3.7 |
| 1939-40 | 10.7 | 14.0 | 13.3 | -9.3 | -9.6 |
| 1940-41 | 9.2 | 9.3 | 10.1 | -4.0 | -4.3 |
| 1941-42 | 7.1 | 12.5 | 6.6 | -1.2 | -2.2 |
| 1942-43 | 5.0 | 10.0 | 3.3 | -1.2 | -2.3 |
| 1943-44 | 5.1 | 1.6 | 2.3 | 3.0 | 4.7 |
| 1944-45 | 4.9 | -0.5 | 1.8 | 6.3 | 4.4 |
| 1945-46 | 7.9 | 4.1 | 4.4 | 10.3 | 10.6 |
| 1946-47 | 3.6 | 7.6 | 5.5 | 3.1 | 1.9 |
| 1947-48 | 5.1 | 9.0 | 7.6 | 0.9 | 0.0 |
| 1948-49 | 2.4 | 4.2 | 2.6 | 1.9 | 1.9 |
| 1949-50 | 2.1 | 4.7 | 2.9 | 2.8 | 1.9 |
| 1950-51 | 8.4 | 9.8 | 9.9 | -1.3 | 0.0 |
| 1951-52 | 8.0 | 8.2 | 1.7 | -0.5 | -1.8 |
| 1939-45 | 7.0 | 7.8 | 6.2 | -1.1 | -1.6 |
| 1945-51 | 4.9 | 6.6 | 5.5 | 3.0 | 2.7 |

*Source:* Calculated from C. H. Feinstein, *National Income, Expenditure and Output of the United Kingdom, 1855-1965* (1972), T22, T42, T140, Cambridge University Press.

could not be sterilised by rising prices. However, in the year following the 1947 sterling crisis it became imperative to contain the growth in consumption in order to direct resources to more urgent needs such as exports. Accordingly, Dalton's last budget in November 1947 reversed one third of his previous tax cuts, the rationing system was tightened up and cuts were imposed on imports and investment. This was the period when bread and potatoes were rationed for the first time and when ration allowances of the basic foods fell below the wartime average. On a weekly basis the average man's allowance of meat per week was 13 ounces, one and half ounces of cheese, six ounces of butter and margarine, one ounce of cooking fat, eight ounces of sugar, two pints of milk and one solitary egg. There was, in addition, a points system for other foodstuffs in short supply, while many luxury imported products such as fruit and canned goods were virtually unobtainable.[55] Housewives spent much time in queuing for food, while black market activities flourished in the age of the 'spiv'. The period was also famous for the appearance of some strange and rather unpalatable foods, including whale-meat steaks, reindeer meet and 'most bizarre of all, snoek, a hitherto unknown, and largely inedible, fish from the warm waters around South Africa which first arrived in May 1948'.[56] These and other schemes, such as the one for producing groundnuts in Tanganyika, left the government open to considerable ridicule.

Despite the severe dietary privations, the nation's health was better than it had ever been thanks largely to the fairer distribution of basic foods together with the additional provisions made for children. 'They were a large and healthy generation, guarded by regulation orange juice, halibut liver oil, and milk.'[57] During the course of 1948 there was some easing of the rationing system, with bread, potatoes and jam being freed. Over the next two years many items were decontrolled so that by 1950 only about 11 per cent of consumers' expenditure was on rationed goods, these being mainly food and coal. Despite his association with austerity, Sir Stafford Cripps left total taxation little changed in his three budgets between 1948-50, while in terms of consumer purchasing power they virtually rescinded Dalton's last act of disinflation.[58] Thus consumption recovered steadily after the severe setback in 1947-48 and it was not until the Korean War crisis that it again fell sharply. By that time rationing was too limited in its coverage to have much effect, and therefore reliance was placed on import cuts and a disinflationary budget in 1951 to rein back consumption. Prices also rose faster than money earnings thus helping to suppress effective demand (Table 5).

Compared with a later period, the relative moderation in consumption helped to free more resources for exports and investment than would otherwise have been the case.[59] It could have been even better had policy been more consistent. The odd thing is that for much of the time fiscal policy was pulling in the opposite direction to physical controls, thus helping to boost consumption. But given the size of the external constraint it could well be argued that consumption should have been more tightly constrained. The external crisis of 1947 was undoubtedly exacerbated by the rapid growth in consumption following the end of the war, while the overall increase between 1945-50 bordered on the excessive given the resource needs elsewhere. Britain had certainly not reached the starvation line, and the working classes were better off than ever before, so in theory more could have been squeezed out. On the other hand, it would undoubtedly have been a difficult course to adopt politically especially as the degree of austerity prevailing in the later 1940s was already causing the government some loss of popularity. Taking account of the severe cutback to consumption during wartime, it was probably the best that could be achieved. After all, even by the turn of the decade the level of consumption *per capita* had only just about recouped the prewar level, whereas in the US during the same period it had risen very considerably. 'After six years of war, the British people were in no mood for heroic sacrifices.'[60]

As for wages and prices there were several factors making for upward pressure in the postwar years. The demand for labour was strong and in several trades severe shortages of manpower prevailed. Union bargaining power had also been strengthened by the increasing density in union membership during the war. At the same time the postwar seller's market lessened the likelihood of employers resisting high wage demands. World shortages meant that there was severe upward pressure on the prices of imported goods, especially food and raw materials which rose considerably faster than the cost of living. Finally, excess purchasing power in Britain gave a further boost to internal prices.

Though domestic prices were rising more rapidly after 1945 than during the latter years of the war, the inflationary potential was not allowed to explode as had been the case after 1918. Extensive price control was retained for several years after the war so that even by 1950 and 1951 some one third of all consumers' expenditure was spent on tightly controlled goods (about the same as in 1946) and a further 10 per cent on loosely price-controlled products (16 per cent in 1946).[61] Furthermore, the rise in the domestic price level was moderated by heavy food and raw-material subsidies. From 1941 through to the beginning of 1949 domestic food prices were virtually insulated from world trends by steadily rising food subsidies which for 1949 were estimated to be running in the region of £568 million. The same was true of many imported materials. However, in the budget of 1949 food subsidies were pegged and then subsequently lowered, while raw-material subsidies were gradually withdrawn. Unfortunately, the change in policy coincided with the import price effect of devaluation in the autumn of that year and later with the commodity price boom following the outbreak of the Korean War. Consequently, retail prices, the increase in which had moderated substantially in the later 1940s, rose very sharply at the turn of the decade.

Price moderation no doubt helped to exacerbate the backlog of frustrated consumer demand, but was probably an essential ingredient in securing the acceptance of wage restraint. The rapid growth in wage rates and earnings following the end of the war was the cause of some concern, and at the beginning of 1947 the government issued a white paper stressing the need to hold down prices and incomes in the interests of international competitiveness.[62] This exhortation met with virtually no response, as might be expected at a time when prices were rising rapidly due to import costs. The government therefore published a somewhat stronger document following the convertiblity crisis. This second white paper,[63] while reaffirming the view that direct interference with the determination of personal incomes was undesirable, argued that there was no justification for any *general* increase in money incomes, including profits, rents and dividends. In effect it was a call for a wages freeze, though with some notable exceptions. Despite some opposition, it was endorsed by the TUC in March 1948 and reaffirmed the following year. Its acceptance by the unions was facilitated by some moderation in import prices after 1947 and the promise of dividend restraint and a once-and-for-all capital levy on former investment income. In 1950, however, the policy began to crumble due to the strong upward pressure in import prices which produced a 10 per cent jump in retail prices the following year. Thus in September 1950 the TUC rejected a continuation of the policy.

While the wage freeze no doubt produced some distortions in the labour market, it did lead to a substantial moderation of wages and earnings.[64] In fact the rate of growth between 1948-50 was only half that of the year preceding the freeze (see Table 5). However, much of the success in restraining incomes can be attributed indirectly to the moderation of price inflation on the import side and the measures to suppress internal price inflation. Once renewed price pressures built up in 1949-51 the basis for an industrial consensus quickly evaporated and wage control was no longer a feasible proposition. Nevertheless, on balance the wage and price record of 1945-51 was better than during the war, as was the growth in the real wage gap. Moreover, despite the increased power of the trade union movement, the period was remarkably free of disputes. In the five years 1945-50 the total number of working days lost in industrial disputes was under 10 million, over half of which were accounted for by transport workers and miners, as against no less than 178 million in the five years 1918-23.[65] Many industries were virtually free of industrial troubles. The continuation of the compulsory arbitration procedure under the wartime Order 1305 of 1940 and the empathy between the Labour Government and the trade union movement no doubt contributed towards keeping the peace.

## Planning, Controls and Economic Policy

In the tight resource situation of the postwar years it could legitimately be argued that investment (especially industrial investment) and exports would have to be the main priorities in terms of claims on resources and that consumption would have to take a back seat. The Labour Government did not deny the importance of these, but it also had several other tasks which it no doubt considered equally urgent. These included full employment, a better distribution of income and more welfare services, the suppression of inflationary pressures, the maintenance of the value of sterling, the containment of debt servicing costs and some relief for the hard-pressed consumer.

Some of these objectives were clearly incompatible and there was no hope of achieving them all at once in the conditions prevailing after 1945. However, until early 1947 the Labour Government had not fully grasped the seriousness of the economic situation and therefore failed to draw up a clear list of priorities. This failing may partly be explained by the fact that things turned out rather better than expected in 1946, that the government was eager to press on with its major policy programme, and that, additionally, it could rely on the system of physical controls to take care of any problems. These factors would also help to explain the government's complacent attitude to overall economic policy which was reflected in the absence of a coherent and consistent macro-policy and a lack of commitment to planning.

In the first year and a half therefore, monetary and fiscal policy were working against the control system since both tended to be expansionary. Dalton, in three successive budgets (1945, 1946, 1947) reduced personal taxes, and the total effect on consumer spending was in the region of 4 per cent of national product.[66] At the same time his drive to bring down both short- and long-term rates of interest undoubtedly exerted an inflationary influence on the economy, though it is easy to exaggerate the contribution of this to the ambitious investment programme of 1946 since several elements were determined by specific policy, for example housing.[67] The motives for cheap money were varied: the desire to reduce the cost of debt servicing including that on international balances, the determination to shift income away from the rentier, and the need to give every encouragement to investment. It should be stressed that the government, with memories of the post-first world war experience, was more concerned with the possibility of too much deflation rather than with the dangers of inflation once the transitional phase had passed.

Whatever the motives for policy, the fact is that both fiscal and monetary policy stimulated consumption and imports unduly, the repercussions of which were to surface in 1947. It seems unlikely that the potentially adverse effects were fully appreciated at the time, though no doubt Labour ministers were confident that their faith in planning and controls would come to the rescue. But if the latter was the case then it is difficult to explain the benign neglect of these matters throughout 1946 and early 1947.

The planning machinery and network of controls inherited from wartime were kept largely intact with little or no attempt to adapt them to peacetime conditions. One or two ministries were wound up, labour controls were disbanded while those over investment were strengthened, but otherwise there were few new initiatives. It could be argued, however, that the wartime system was totally unsuited to the varied needs of peacetime. Once the pressure of war needs was removed, there was no longer a single criterion to determine the allocation of resources, while the absence of any overall plan and the waning interest of departments in resource budgeting meant that there were no clear principles on which to proceed. Moreover, there was a distinct lack of coordination at the centre, for the Lord President's Committee, which in wartime had been the supreme coordinating body in economic matters, proved to be, under Herbert Morrison, singularly ill-equipped for the task in peacetime. The Committee did not exercise any powers that could be regarded as planning, nor did the Lord President himself, who was already busily engaged in supervising the government's massive legislative programme, show a great deal of enthusiasm for the subject.[68] At the same time the Treasury, which was still in the subordinate role to which it had been relegated in wartime, was even less inclined to offer initiatives in the field of planning, especially as its incumbent was more at home in the world of finance than in those of planning and Keynesian economics.

The results of the low priority accorded to planning and control, surprising though that may appear given Labour's election promise 'to plan from the ground up', are readily identifiable. During 1946 and early 1947 there was an enormous overissue of authorisations and allocations by sponsoring departments with very little central coordination or direction, the worst excesses occurring in the building industry.[69] Secondly, there was no attempt to develop a coherent policy for dealing with the impending coal shortage, which had been anticipated as far back as May 1946 partly as a result of the difficulty of recruiting workers to the industry. Nor did the Lord President's Committee spare much thought as to the possible repercussions on sterling and the balance of payments likely to arise with the approach of the scheduled date for sterling convertibility in July 1947.

Thus to begin with the Labour Government had no coherent planning strategy. The government, judging by the few references which were made to the subject, appeared to regard planning as of secondary importance, and few ministers, apart perhaps from Shinwell, had any clearly formulated views on the subject. Morrison himself set the tone of the debate when he stated in the autumn of 1946 that 'after all planning, though big and complicated, is not much more than applied commonsense'.[70] For the most part therefore the government was content to rely on using the wartime physical and financial controls to allocate resources on an *ad hoc* basis rather than according to the requirements of any published plan.

The government was rudely shattered out of its complacency in 1947, the year when everything turned sour, and one eventful in terms of economic policy. A winter fuel shortage and a summer sterling crisis led to a tightening of controls, an overhaul of the planning machinery, a deflationary autumn budget, and finally to a change of Chancellor following a budgetary indiscretion by Dalton. It also saw the publication of what purported to be Labour's first planning document. The latter was not however a response to the crises since it had been foreshadowed in November 1946 and subsequently appeared towards the end of the fuel crisis.

The first *Economic Survey*, published in February 1947,[71] is worth looking at briefly since it is the nearest the government got to formulating a planning exercise.[72] It was prepared by the Lord President's staff and contained three main sections, on economic planning, a review of economic events and problems in 1945-46, and an extended account of expected developments in 1947. As to general principles, it recognised the need to plan in order to use 'the national resources in the best interests of the nation as a whole', but the emphasis was clearly on the short term and conditioned by immediate bottlenecks such as foreign exchange, manpower, fuel supplies and raw-material requirements. 'It is too early yet to formulate the national needs over, say, a five year period with enough precision to permit the announcement of a plan in sufficient detail to be a useful practical guide to industry and the public. There are still too many major uncertainties, especially in the international field.'[73] This is a curious confession, almost an admission that long-term planning should be reserved for good times, and somewhat gratuitous in view of the fact that the Lord President's Committee was known to be working on a longer-term programme. The document also acknowledged that in a democratic society there were limits as to how far one could go with planning and control. It would have to be 'planning by persuasion' rather than totalitarian control; development could be influenced to an extent by physical and financial controls but in the final analysis the real effectiveness of any plan would depend on the 'combined effort of the whole people'. Clearly the authorities had doubts as to what could be achieved in a free society.

The *Survey*, according to Joan Mitchell, did constitute a plan of sorts since it presented quantitative estimates and prepared economic budgets setting out requirements and resources in terms of both manpower and national income and expenditure.[74] The resources table only showed the percentage distribution of output among main sectors and no attempt was made to estimate the total increment to output. Nevertheless, the budgets were essential for allocation purposes since they would reveal the gaps or imbalances that had to be rectified. But on the crucial question of how balance was to be brought about the *Survey* was less sanguine. It recognised that the control apparatus at the disposal of the government could not bring about very rapid changes or make fine adjustments.[75] Controls, as Dow was later to observe, were far more effective in sharing out what supplies were actually available, rather than in altering the quantities in question.[76] For this reason therefore, success would depend very much on the cooperative effort of both sides of industry and of the nation as a whole.[77]

The *Survey* cannot therefore be regarded as a planning blueprint since much of it reads like a statement of hope rather than one of firm intention. As Kaldor noted at the time, it was 'more in the nature of what is likely to happen if things go reasonably well than of intermediary objectives in a clearly conceived long-term plan.'[78] That its targets or forecasts were not fulfilled is hardly surprising since its projections were rapidly overtaken by the course of events, though this failure did raise doubts as to whether it was possible to plan at all given the type of machinery and methods then being used.

The publication of the *Survey* was followed by a series of changes in economic machinery and organisation designed to provide stronger central direction of the economy and strengthen the inter-departmental planning organisation. Some of these were foreshadowed by Sir Stafford Cripps (President of the Board of Trade), when introducing the debate on the *Survey* early in March. Towards the end of that month the Prime Minister announced the setting up of the Central Economic Planning Staff (CEPS) and the Economic Planning Board (EPB). The former was an interdepartmental body chaired by the Chief Planning Officer (Sir Edwin Plowden). Its main tasks were those of drawing up a long-term plan for the use of the country's resources, monitoring the progress of the *Survey* and keeping in touch with all the relevant departments. The Chief Planning Officer was to work directly under the Lord President and the group had no independent executive powers since all decisions on planning would ultimately be taken by the Cabinet. Thus the CEPS was what we might now call a 'think tank' for advising the government on strategy and policy with particular reference to planning.

The Economic Planning Board did not in fact materialise until July when its membership and functions were announced. It was headed by the Planning Officer and consisted of permanent secretaries of various ministries, representatives from employers' bodies, the TUC and the CEPS. It too had purely advisory functions, namely those of advising the government on the best ways of using economic resources, both for the realisation of long-term aims and for remedial purposes relative to immediate needs.

Following the convertibility crisis the Prime Minister announced yet further changes (early August) pursuant to a new drive to expand exports and more import cuts. The upshot was the establishment of a new Ministry of Economic Affairs with a small staff under Cripps, which would take over the CEPS and the Economic Information Unit of the Treasury. It also assumed the coordinating responsibility of the Lord President's Committee which was subsequently disbanded and replaced by two new ones: the Economic Policy Committee, chaired by the Prime Minister and including the Chancellor, the Minister of Economic Affairs and the Lord President, to deal with major issues of internal and external economic policy, and the Ministerial Steering Committee, under Cripps, to look after production and the day-to-day details of the economic programme.

Finally, on 13 November, Cripps replaced Dalton as Chancellor following his budget leak, taking with him responsibilities for the general coordination of economic policy. The CEPS, the Economic Planning Board, the Chief Planning Officer and the Economic Information Unit now became part of the Treasury.[79] The latter therefore regained its former status and became the central economic and planning department with Cripps in a very dominant position.

Thus it had proved a very arduous and cumbersome process of reform. Many of the changes were more cosmetic than real and had little immediate impact. In fact it was really not until Cripps assumed the role of economic overlord that the central direction of economic policy formation was visibly strengthened despite all the intervening tinkering with the machinery. Under Cripps as Chancellor there is certainly more evidence of firmer control and direction, though it is doubtful whether the new machinery made all that much difference to policy in practice, nor did it lead to a more rigorous planning effort.

In fact, as far as planning is concerned, the evidence suggests that from about the middle of 1948 interest in the subject began to wane. Certainly Britain never had in this period what might be called a proper plan 'if by economic plan is meant a set of economic objectives, integrated and consistent in their assumptions, which the Government had decided to carry out and which they had the power to carry out'.[80] It is true that the resource budgeting approach was pursued more vigorously by the Planning Board under Cripps than had been the case hitherto, while the central coordination of economic policy was improved after 1947. In addition, the *Economic Surveys* continued to appear annually and the Planning Board and the CEPS eventually produced a long-term programme which was required by OEEC under the Marshall Plan.

As regards the former, the 1948 *Survey* was probably a slightly more sophisticated document than its predecessor, but it clearly took a jaundiced view of the role of the planner: 'Many of the factors governing its [Britain's] economic development are quite beyond the effective influence of the planner.'[81] But whereas the first two *Surveys* did at least deal with planning and targets, the later ones had much less to say on these matters; their language became more conservative and cautious and they tended to concentrate on the past and the present rather than the future. As *The Economist* remarked: 'The perplexing thing about the *Survey* for 1950 is its lack of plan.'[82] Catterall has suggested that the *Surveys* really became complementary to the long-term plan in the sense of progress reports on that programme, and that for this reason they ceased to be planning documents in any sense of the term.[83] That may be so, but the long-term programme had a short active life. It was published as a white paper in December 1948,[84] but it was very soon quietly forgotten by ministers and officials alike. It therefore cannot be said to have served as a continuing frame of reference to the authorities.[85] It was not a detailed forecast of what might happen, but it did contain several targets for industrial sectors and the balance of payments, the equilibrium of which was seen as the main objective. It also contained a suggested distribution of investment for major sectors of activity. Whether the programme was anything more than a good forecast of the likely outcome from using controls reasonably efficiently, rather than a plan guiding the use of controls and putting forward ambitious but unrealistic targets, is not readily apparent, according to Mitchell.[86] However, since the programme was shelved away to gather dust it is doubtful whether it could have had very much influence on economic policy.

In the absence of a coherent planning framework Cripps used a combination of policy weapons, but with greater internal consistency than had been the case under his predecessor. Monetary policy, after the collapse of Dalton's cheap money experiment, was kept basically neutral with care being taken to see that it did not work against the general aim of disinflation, as had been the case under Dalton.[87] Cripps tried hard to disinflate by fiscal means though in fact his three budgets (1948-50) reduced taxes on consumption which more or less offset Dalton's last budget. It was left to Gaitskell, under the pressure of rearmament, to produce the biggest rise in taxes since the war in his defence budget of 1951.[88] Taking the period as a whole (1945-51), fiscal policy was passive rather than actively disinflationary since taxes on balance were not raised but simply kept high. The economy therefore slowly disinflated of its own accord through higher saving and a gradual easing in supply conditions.[89]

Physical controls continued to be important despite their gradual removal and the tendency for their effectiveness to diminish over time. They were still quite extensive by 1950, with 73 per cent of all imports being controlled in one way or another, 47 per cent of raw-material inputs into industry subject to allocation, and 45 per cent of consumers' expenditure going on price-controlled goods. Rationing however accounted for only about 11 per cent of consumer spending, mainly basic foods and coal. In addition, there was still fairly extensive control over investment by one means or another. In practice controls were subject to alternate tightening and slackening, especially those over investment and imports, as conditions dictated. Cripps also resorted to verbal persuasion and exhortation in an effort to gain the cooperation of employers and workers in promoting the national effort. His two major successes in this respect were getting industry to respond to the export drive and persuading the unions to accept a wage freeze. On the other hand, the government failed in its attempt to get private industry to establish tripartite development councils to improve efficiency and organisation under the Industrial Organisation and Development Act of 1947.

Why, one may ask, was the exercise in planning so limited and so timid given the Labour Party's original election promise on this matter? There are several possible answers to this question. For one thing the government delayed too long before it launched its first tentative planning document, partly because of the inadequate machinery for getting things moving in this respect. Its publication came at the worst possible time, in the midst of the fuel crisis, to be followed later in the year by the sterling crisis, both of which strained the credibility of the document. Labour in fact never fully recovered from the traumas of 1947 which shook the faith of those who thought it could plan, largely because it had done so little to anticipate the difficulties of that year.

Secondly, the whole machinery of government was overloaded in the first eighteen months or so of office. At the beginning things had gone relatively smoothly for the government with the result that it overreached itself. The Prime Minister tacitly admitted as much in the House of Commons debate on economic policy in the summer of 1947 when he acknowledged that the government's problems were caused by having tried to do too much too quickly.[90] During this period ministers were too busy steering major welfare and nationalisation measures through Parliament to give much attention to planning and the coordination of economic policy. In the fuel crisis, for example, Shinwell was much preoccupied with piloting the Electricity Nationalisation Bill through the House. Subsequently, when attention was turned to the question of reforming the economic machinery of government it took the best part of 1947 before the reorganisation was complete and even then it is doubtful whether it was all that effective in terms of planning the economy or coordinating the direction of economic policy. Certainly until the end of 1947 there was too much fragmentation in the direction of economic matters.[91]

By the time the new machinery was in place the government had effectively lost the initiative, and probably some of its enthusiasm for a planned society. Cripps, as Chancellor, was also known to be more in favour of using persuasion rather than directives to achieve his objectives. Moreover, not only had the government experienced a series of unfortunate mishaps, but it also found to its dismay that it was not quite so easy in peacetime to achieve forecasts or targets by using controls, simply because there was now more than a single criterion for determining the allocation of resources. For example, the limited success which attended its efforts to direct manpower into essential industries under the revived Control of Engagement Order of October 1947, demonstrated that its ability to bring about desired changes in resource use was limited, short of draconian measures.

That the control machinery worked less effectively in peacetime than during the war there can be no doubt. The diversity and complexity of peacetime production and demand meant that it required constant adjustment to changing conditions. Unfortunately this proved to be the weakest aspect of the whole system. As the controlling departments lost part of their wartime staff the administrative machinery became less capable of coping with the burden: 'the sheer weight of day-to-day work imposed on the Civil Service by the various systems of controls makes it difficult in some Departments both to keep the machinery running and at the same time keep it in view as a whole, and carry out overhauls.'[92] The small staffs employed in some of the planning agencies inhibited any improvement being made in the control apparatus and rendered impossible the provision of adequate information on such things as stocks and work in progress which were vital necessities for its success. It is not surprising therefore that at times the control machinery tended to produce unsatisfactory results, a situation not helped by the fact that the control network became less comprehensive with the passage of time.[93] The difficulties encountered at the micro level have been well demonstrated by Rosenberg's detailed study of the building industry in which he shows how the government's ambitious building programme was thwarted by the restricted scope of the building controls and the limited nature of the administrative apparatus of enforcement.[94]

Finally, one must remember that controls had to work in an increasingly hostile environment in the postwar years. They could not always be enforced by statutory rules alone and in some cases voluntary control was necessary. Initially, the community gave its acquiesence to the need for controls. But as economic conditions gradually improved, industrialists and traders became less willing to cooperate with the government in their administration. Even trade unionists, who tended to be strong advocates of planning and controls, objected when these impinged upon their own interests.[95] The business world and the public at large were less prepared to accept continuing austerity and restriction and became increasingly frustrated and annoyed by the excessive detail of control and by what they considered to be the misdirection of resources under planning.[96] In response to mounting criticism (including that of the press and the Tory Opposition), the Board of Trade had two bonfires of controls in 1948-49. Though the control apparatus still remained extensive, its gradual dismantlement diminished the authorities' ability to steer the economy by physical means, leaving them no alternative but to resort increasingly to more indirect techniques such as fiscal and monetary policy. Moreover, the failure of the government to reach an accord with employers on its industrial policy also spelled the end of any hopes of planning.[97] It was in these changing conditions that planning and controls were coming to be viewed, both by the public and the government 'more as necessary evils to be dispensed with as quickly as possible than as the indispensible machinery for Socialist policy'.[98]

## Labour's Record

By the turn of the decade the Labour Government had virtually run out of steam. The massive legislative programme had taken its toll on the ministerial team, while its popularity in the country was declining due partly to continued austerity and rationing. In the general election of February 1950 the government was returned but with a tiny majority and as a result it had few new initiatives. In the following year the Conservatives were returned to power.

An overall assessment of Labour's economic and social achievements is not an easy task since there are so many angles from which the record may be judged. In terms of reconstruction and productive effort it did quite well given the periodic crises on the external side. The unwinding of the war machine was accomplished fairly smoothly and with no repetition of the boom and bust episode as in 1919-21. Growth rates were very respectable compared with later years and even more so when set against the poor performance following the first world war. Exports were particularly strong, growing at a compound rate of 11.9 per cent a year between 1946-51,[99] while industrial production averaged 5¾ per cent. Total output (GDP) rose more modestly, 2.3 per cent a year, largely because the service sector was held back. The unemployment record was excellent, falling from 3.1 per cent in 1947 to 1.2 per cent by 1951. Even the balance of payments showed signs of stabilising by 1950, though the dollar problem was far from solved, while in the following year the Korean War upset things again. Compared with Western Europe Britain's performance was commendable, though in the latter part of the period there were signs that other countries were beginning to pull ahead of Britain.

Labour carried through a remarkable legislative programme, especially in the field of nationalisation and welfare. Yet big as this was it did not create a social revolution. There were no fundamental changes in the structure of society.[100] Little modification occurred in the class system nor was there any shift in the power base. No major transfer of power took place in the financial, commercial and industrial establishment and the civil service was left intact. Ownership might change but power did not; the workers' role remained unaltered and the unions played no part in management, nor did they wish to. There was no significant redistribution of wealth, and very little of income for that matter. A very small proportion of the population continued to own most of the wealth and property of the country. Tax and transfer payments led to a slight shift in income towards the lower income groups, though probably less so than in wartime, but real poverty remained despite the welfare state. For the mass of the population real wage gains were very modest (less than in wartime), while rationing and shortages continued to depress consumption.

As far as the general management of the economy is concerned, one may point to several credits. Inflationary pressures were contained, albeit at times by cosmetic devices such as subsidies, and resources were guided into exports and investment. Full employment was achieved and debt service costs were contained by low interest rates which diverted income away from the rentier. Within their obvious limitations physical controls were used successfully to produce a fairer distribution of scarce resources and to guide these into priority channels, though not all of them worked, for example direction of labour. On the other hand, policy sometimes lacked consistency, while efforts to plan the economy were, to put it mildly, a fiasco. There was never any real commitment to planning, despite frequent utterances to the contrary, and the vain hope of using controls to support planning eventually became impractical as the pressure for their removal mounted. The reform of the planning and control machinery in 1947 was more cosmetic than real and did not denote any significant change in direction.

Did the Labour Government try to do too much in too short a space of time? Lewis once argued that it spent too much effort, for ideological reasons, on public ownership and welfare measures to the detriment of other matters.[101] There have of course been frequent criticisms to the effect that the administrative machine was overloaded and, as we have seen, Attlee himself admitted as much in 1947. In some respects there is substance in these views insofar as attention was deflected from more immediate problems, the fuel crisis of 1947 being a good case in point. Several contemporary economists also believed that the investment programme was over-ambitious, and at the time of the sterling crisis in 1947 Harrod, Meade, Lewis and Jewkes were instrumental in persuading the government to trim projected investment. Conversely, later commentators, notably Shonfield, have criticised the administration for failing to assign a high enough proportion of national income to investment. 'The Labour Party leaders, especially Sir Stafford Cripps, were peculiarly complacent over this whole issue. Partly it was because they got the relevant figures consistently wrong, in a sense which allowed them to over-estimate their own achievement.'[102] He appears happier with the distribution of investment in that a larger share was directed to plant and machinery, though not everyone would be satisfied on this score. It could be argued that too much went into housing and welfare at the expense of sectors such as coal, railways and electricity, and that industry bore the brunt of the cuts in projected investment in 1947, 1949 and 1951. Ultimately of course, it all depends on one's own particular judgement as to the order of priorities. One may say however that given the constraints the overall investment record was satisfactory. Gross domestic fixed capital formation had surpassed the prewar level by 1948, the gross and net capital stocks had been restored, while the investment ratio was better than it had ever been.

Finally, the drive for exports undoubtedly paid off in quantum terms to 1950 and the product mix was changing for the better. But there was no comparable shift in market structure. The Labour Government turned its back on Europe and British exporters failed to benefit from the revival of the West European economies. Imperial connections died hard and in this respect, as in many other matters, Labour appeared no less conservative than the Tory Opposition or the public at large.

From the long-term point of view, perhaps the most signal failure of the Attlee Administration was its inability to give a new sense of purpose and direction on economic matters to the British public. Growth, productivity, technological change and the need to export were much discussed by ministers and officials but they never became part of the working vocabulary of the average citizen. Little attempt was made to show how these could contribute to real income gains and economic well-being and hence the British people were left to surmise these were of no direct relevance to their life. This failure in the learning process left British citizens ill-equipped to face the more competitive and challenging climate of later decades.

## Notes

1. K.O. Morgan, *Labour in Power 1945-1951* (1984), pp. 35, 41.
2. R.B. McCullum and A. Readman, *The British General Election of 1945* (1947), p. 44.
3. The development of Labour's thinking on public ownership and other economic matters is explored at length in E. Durbin, *New Jerusalems: the Labour Party and the Economics of Democratic Socialism* (1985).
4. Quoted in Morgan, *op. cit.*, p. 111.
5. *Let us Face the Future* (1945), p.4.
6. C.R. Attlee, *Purpose and Policy: Selected Speeches* (1947), p. 23.
7. *48th Annual Conference Report of the Labour Party*, June 1949, p. 154.
8. Morgan, *op. cit.*, p. 24.
9. *Ibid.*, p. 22; J.C.R. Dow, *The Management of the British Economy, 1945-60* (1964), p. 11.
10. See E. Zupnick, *Britain's Postwar Dollar Problem* (1957).
11. Details of the Loan Agreement can be found in R.N. Gardner, *Sterling Dollar Diplomacy* (1956), Chapter 10 and pp. 213-21.
12. A.S. Milward, *The Reconstruction of Western Europe, 1945-51* (1984), p. 41.
13. Dow reckons that it accounted for only one fifth of the worsening dollar outflow in that year. Dow, *op. cit.*, p. 24.
14. A.R. Conan, *The Problem of Sterling* (1966), p. 28.
15. C.C.S. Newton, 'The Sterling Crisis of 1947 and the British Response to the Marshall Plan', *Economic History Review*, 37 (1984), pp. 397-8.
16. In April 1948 it became the Organisation for European Economic Co-operation (OEEC).
17. Milward, *op. cit.*, pp. 94-5.
18. W.C. Mallalieu, *British Reconstruction and American Policy 1945-1955* (1956), p. 71.
19. Milward, *op. cit.*, p. 100.
20. A. Cairncross and B. Eichengreen, *Sterling in Decline* (1983), p. 144.
21. Cairncross and Eichengreen, *op. cit.*, p. 147.
22. Conan, *op. cit.*, pp. 29-30.
23. Cairncross and Eichengreen, *op. cit.*, p. 151.
24. Dow, *op. cit.*, p. 47.
25. Cairncross and Eichengreen, *op. cit.*, pp. 151-5; *Economic Survey for 1950*, Cmd. 7914 (1950), p.6.
26. D.H. Aldcroft, 'The Effectiveness of Direct Controls in the British Economy, 1946-1950', *Scottish Journal of Political Economy*, 10 (1963), pp. 239-40.
27. Conan, *op. cit.*, p. 55.
28. Milward, *op. cit.*, pp. 456, 457.
29. Conan, *op. cit.*, pp. 8-9; S. Pollard, *The Development of the British Economy, 1914-1980* (1983), p. 241.
30. Economic Cooperation Administration, *The Sterling Area* (1951), p. 75, quoted in Eichengreen and Cairncross, *op. cit.*, p. 144.
31. GDP data are given in A. Maddison, *Phases of Capitalist Development* (1982), pp. 174-7.
32. G.D.N. Worswick and P.H. Ady, *The British Economy, 1945-1950* (1952), p. 393.

33. Mallalieu, *op. cit.*, p. 155.
34. United Nations, *Economic Survey of Europe in 1951*, p. 179.
35. Zupnick, *op. cit.*, p. 69.
36. Worswick and Ady, *op. cit.*, p. 70.
37. Milward, *op. cit.*, p. 28.
38. Cf. Morgan, *op. cit.*, pp. 391-2, 492, 493, who tends to overemphasise the rigidity of the industrial structure and hence presumably that of exports.
39. Milward, *op. cit.*, p. 339.
40. A.A. Rogow and P. Shore, *The Labour Government and British Industry, 1945-1951* (1955), p. 170.
41. Later merged with BOAC.
42. B.W. Lewis, *British Planning and Nationalization* (1952), p. 111.
43. K.M. Gwilliam, *Transport and Public Policy* (1964), p. 102.
44. Morgan, *op. cit.*, p. 187.
45. H. Pelling, *The Labour Governments, 1945-51* (1984), p. 97; J. Stevenson, *British Society, 1914-45* (1984), pp. 454-7; A.J. Peacock and J. Wiseman, *The Growth of Public Expenditure in the United Kingdom* (1961), p. 94.
46. Family allowances had been implemented by legislation introduced by the Churchill Caretaker Government of 1945.
47. Pelling, *op. cit.*, p. 110; J.R. Short, *Housing in Britain* (1982), pp. 42, 45.
48. Feinstein, *op. cit.*, T86.
49. Milward, *op. cit.*, p. 479.
50. Morgan, *op. cit.*, p. 177.
51. Peacock and Wiseman, *op. cit.*, p. 84. Social services cover education and child care, health services, national insurance (unemployment, sickness, retirement pensions), national assistance, family allowances, housing (subsidies and capital spending) and food subsidies.
52. *Ibid.*, pp. 187, 191.
53. Pollard, *op. cit.*, p. 271; Worswick and Ady, *op. cit.*, p. 371.
54. Dow, *op. cit.*, p. 201.
55. Morgan, *op. cit.*, pp. 369-70; M. Sissons and P. French (eds.), *Age of Austerity, 1945-51* (1963), p. 38.
56. Morgan, *op. cit.*, pp. 369-70.
57. Sissons and French, *op. cit.*, p. 37.
58. Dow, *op. cit.*, p. 201.
59. Zupnick, *op. cit.*, pp. 58-9.
60. Sissons and French, *op. cit.*, p. 187.
61. Dow, *op. cit.*, p. 176.
62. *Statement on the Economic Considerations affecting Relations between Employers and Workers*, Cmd. 7018 (1947).
63. *Statement on Personal Incomes, Costs and Prices*, Cmd. 7321 (1948).
64. B.C. Roberts, *Wages Policy in War and Peace* (1958).
65. Morgan, *op. cit.*, p. 499.
66. Dow, *op. cit.*, p. 201.
67. Worswick and Ady, *op. cit.*, p. 194.
68. J. Mitchell, *Groundwork to Economic Planning* (1966), p. 62.
69. Dow, *op. cit.*, p. 151.
70. J. Leruez, *Economic Planning and Politics in Britain* (1975), p. 38.
71. *Economic Survey for 1947*, Cmd. 7046 (1947).
72. That is if we discount the long-term programme prepared for OEEC (see below).
73. *Economic Survey for 1947*, para. 14.

74. Mitchell, *op. cit.*, p. 66.
75. Leruez, *op. cit.*, pp. 47-8.
76. Dow, *op. cit.*, p. 16.
77. *Economic Survey for 1947*, para. 27.
78. *The Times*, 25 February 1947.
79. The post of Minister of Economic Affairs was combined with the Treasury. After the election of February 1950 it was revived and held by Hugh Gaitskell until 19 October 1950, to assist the ailing Cripps.
80. D.N. Chester in Worswick and Ady, *op. cit.*, p. 360.
81. *Economic Survey for 1948*, Cmd. 7344 (1948), p. 56.
82. *The Economist*, 1 April 1950, p. 689.
83. R.E. Catterall, *Economic Planning in Britain: An Historical Study*, B.A. thesis, University of Leicester, 1970, p. 34.
84. *European Cooperation: Memorandum Submitted to the OEEC Relating to Economic Affairs in the Period 1949 to 1953*, Cmd. 7572 (1948).
85. Leruez, *op. cit.*, p. 57.
86. Mitchell, *op. cit.*, p. 117 *et seq.*
87. Dow, *op. cit.*, p. 228.
88. *Ibid.*, p. 201.
89. *Ibid.*, p. 210.
90. Leruez, *op. cit.*, pp. 45-6.
91. A further problem noted by Chester was the dilemma raised between coordination and decentralised administration. 'The more one tries to co-ordinate everything and fit the economic details into a single plan the more one puts an impossible load on the central machinery. On the other hand, the more one pushes off the decisions and detail on to the departments and other bodies the less chance that the details will fit together.' D.N. Chester in Worswick and Ady, *op. cit.*, pp. 362-3.
92. *Report of the Committee on Intermediaries*, Cmd. 7904 (1950), para. 99.
93. D.H. Aldcroft, 'The Effectiveness of Direct Controls in the British Economy, 1946-1950', *Scottish Journal of Political Economy*, 10 (1963), p. 228.
94. N. Rosenberg, *Economic Planning and the British Building Industry, 1945-49* (1960), pp. 130, 150.
95. *79th Annual Report of the TUC*, (1947), p. 354.
96. In the year ending March 1949, no less than 19 million applications were made to government departments for licences, permits, etc. quite apart from the issue of ration books. Cmd. 7904 (1950), para. 8. The *Daily Dispatch* (21 January 1949) reported the case of one poor gentleman who was asked to produce a certificate of proof that his leg was still amputated before he could claim his extra soap ration.
97. Leruez, *op. cit.*, p. 72.
98. A.A. Rogow and P. Shore, *The Labour Government and British Industry, 1945-51* (1955), p. 47.
99. No less than 15.2 per cent a year between 1946-50.
100. R.A. Brady, *Crisis in Britain* (1950), p. 661.
101. W.A. Lewis, *The Principles of Economic Planning* (1963), p. 113.
102. A. Shonfield, *British Economic Policy since the War* (1958), p. 174.

# Select Reading

The literature on the twentieth century is now so vast that a comprehensive bibliography would add at least another chapter to this volume. In any case a lengthy bibliography might do more to confuse than enlighten the reader who is starting out on a study of this period. I have therefore confined my selection to a number of more general works to supplement the material presented in the text, and anyone wishing to dig deeper may follow up the footnote references to each chapter.

The two most useful general works covering all aspects of British history are C.L. Mowat, *Britain between the Wars, 1918-1940* (Methuen, 1955) and K.O. Morgan, *Labour in Power, 1945-1951* (Oxford University Press, 1984). Mowat's work, though somewhat dated, is still far the best general survey of the interwar period. Morgan's study of the postwar period should be read in conjunction with Sir Alec Cairncross's survey of economic policy, *Years of Recovery: British Economic Policy 1945-51* (Methuen, 1985).

Economic surveys covering the whole of the period can be found in S. Pollard, *The Development of the British Economy, 1914-1980* (Edward Arnold, 1983), which is a mine of useful information, and A.J. Youngson, *The British Economy, 1920-1957* (Allen and Unwin, 1960), which is a shorter and selective treatment of the main issues. A brief survey is also provided by M.W. Kirby, *The Decline of British Economic Power since 1870* (Allen and Unwin, 1981). The social side is well presented in J. Stevenson, *British Society, 1914-45* (Penguin, 1984). A much more advanced analysis of growth covering a longer period may be found in R.C.O. Matthews, C.H. Feinstein and J.C. Odling-Smee, *British Economic Growth, 1856-1973* (Stanford University Press, 1982).

Readers wishing to familiarise themselves with the international and European backgrounds may like to consult the following: D.H. Aldcroft, *The European Economy, 1914-1980* (Croom Helm, 1980), and two volumes by A.S. Milward, *War, Economy and Society, 1939-1945* (Allen Lane, 1977) and *The Reconstruction of Western Europe, 1945-51* (Methuen, 1984).

# Political Glossary — Chief Economic Ministers

*Coalition Government January 1919 - October 1922*

| | |
|---|---|
| Prime Minister | David Lloyd George |
| Chancellor of Exchequer | A. Chamberlain<br>Sir R. Horne (1 April 1921) |
| Board of Trade | Sir A. Stanley<br>Sir A. Geddes (26 May 1919)<br>Sir R. Horne (19 March 1920)<br>S. Baldwin (1 April 1921) |
| Board of Agriculture and Fisheries (Ministry 15 August 1919) | R.E. Prothero<br>Lord Lee (15 August 1919)<br>Sir A. Griffith-Boscawen (13 February 1921) |
| Ministry of Labour | Sir R. Horne<br>T. Macnamara (19 March 1920) |
| Ministry of Transport | Sir A. Geddes<br>Viscount Peel (7 November 1921)<br>Earl of Crawford (12 April 1922) |

*Conservative Government October 1922 — May 1923*

| | |
|---|---|
| Prime Minister | A. Bonar Law |
| Chancellor of Exchequer | S. Baldwin |
| Board of Trade | Sir P. Lloyd-Greame |
| Ministry Agriculture and Fisheries | Sir R. Saunders |
| Ministry of Labour | Sir A. Montague-Barlow |
| Ministry of Transport | Sir J. Baird |

*Conservative Government May 1923 — January 1924*

| | |
|---|---|
| Prime Minister | Stanley Baldwin |
| Chancellor of Exchequer | N. Chamberlain (27 August 1923) |
| Board of Trade | Sir P. Lloyd-Greame |
| Ministry Agriculture and Fisheries | Sir R. Saunders |
| Ministry of Labour | Sir A. Montague-Barlow |
| Ministry of Transport | Sir J. Baird |

*Labour Government January — November 1924*

| | |
|---|---|
| Prime Minister | J. Ramsay MacDonald |
| Chancellor of Exchequer | P. Snowden |
| Board of Trade | S. Webb |
| Ministry Agriculture and Fisheries | N. Buxton |
| Ministry of Labour | T. Shaw |
| Ministry of Transport | H. Gosling |

*Conservative Government November 1924 — June 1929*

| | |
|---|---|
| Prime Minister | Stanley Baldwin |
| Chancellor of Exchequer | W.S. Churchill |
| Board of Trade | Sir P. Cunliffe-Lister (formerly Lloyd-Greame) |
| Ministry Agriculture and Fisheries | E.F.L. Wood<br>W. Guiness (4 November 1925) |
| Ministry of Labour | Sir A. Steel-Maitland |
| Ministry of Transport | W. Ashley |

*Labour Government June 1929 — August 1931*

| | |
|---|---|
| Prime Minister | J. Ramsay MacDonald |
| Chancellor of Exchequer | P. Snowden |
| Board of Trade | W. Graham |
| Ministry Agriculture and Fisheries | N. Buxton<br>C. Addison (5 June 1930) |
| Ministry of Labour | Margaret Bondfield |
| Ministry of Transport | H. Morrison |

*National Government August — November 1931*

| | |
|---|---|
| Prime Minister | J. Ramsay MacDonald |
| Chancellor of Exchequer | P. Snowden |
| Board of Trade | Sir P. Cunliffe-Lister |

*National Government November 1931 — June 1935*

| | |
|---|---|
| Prime Minister | J. Ramsay MacDonald |
| Chancellor of Exchequer | N. Chamberlain |
| Board of Trade | W. Runciman |
| Ministry Agriculture and Fisheries | Sir J. Gilmour<br>W. Elliot (28 September 1932) |
| Ministry of Labour | Sir H. Betterton<br>O. Stanley (29 June 1934) |
| Ministry of Transport | J. Pybus<br>O. Stanley (22 February 1932)<br>L. Hore-Belisha (29 June 1934) |

*National Governments June 1935 — May 1940*

| | |
|---|---|
| Prime Minister | Stanley Baldwin<br>Neville Chamberlain (28 May 1937) |
| Chancellor of Exchequer | N. Chamberlain<br>Sir J. Simon (28 May 1937) |
| Board of Trade | W. Runciman<br>O. Stanley (28 May 1937)<br>Sir A. Duncan (5 January 1940) |
| Ministry Agriculture and Fisheries | W. Elliot<br>W. Morrison (29 October 1936)<br>Sir R. Dorman Smith (29 January 1939) |
| Ministry of Labour (Ministry of Labour and National Service 3 September 1939) | E. Brown |
| Ministry of Transport | L. Hore-Belisha<br>L. Burgin (28 May 1937)<br>E. Wallace (21 April 1939) |
| Ministry of Food | Lord Woolton (3 April 1940) |
| Ministry of Supply | L. Burgin (14 July 1939) |

*Coalition Government May 1940 — May 1945*

| | |
|---|---|
| Prime Minister | Winston S. Churchill |
| Chancellor of Exchequer | Sir Kingsley Wood |
| | Sir J. Anderson (24 September 1943) |
| Board of Trade | Sir A. Duncan |
| | O. Lyttelton (3 October 1940) |
| | Sir A. Duncan (29 June 1941) |
| | J. Llewellin (4 February 1942) |
| | H. Dalton (22 February 1942) |
| Ministry Agriculture and Fisheries | R. Hudson |
| Ministry of Labour and Nat. Service | E. Bevan |
| Ministry of Transport | Sir J. Keith |
| (Ministry of War Transport 1 May 1941) | J. Moore-Brabazon (3 October 1940) |
| | Lord Leathers (1 May 1941) |
| Ministry of Food | Lord Woolton |
| | J. Llewellin (11 November 1943) |
| Ministry of Fuel Light and Power | G. Lloyd-George (3 June 1942) |
| Ministry of Supply | H. Morrison |
| | Sir A. Duncan (3 October 1940) |
| | Lord Beaverbrook (29 June 1941) |
| | Sir A. Duncan (4 February 1942) |

*Caretaker Government May — July 1945*

| | |
|---|---|
| Prime Minister | Winston S. Churchill |
| Chancellor of Exchequer | Sir J. Anderson |
| Board of Trade | O. Lyttelton |
| Ministry Agriculture and Fisheries | R. Hudson |
| Ministry of Labour and Nat. Service | R. Butler |
| Ministry of War Transport | Lord Leathers |
| Ministry of Food | J. Llewellin |
| Ministry of Fuel and Power | G. Lloyd-George |
| Ministry of Supply | Sir A. Duncan |

*Labour Governments July 1945 — October 1951*

| | |
|---|---|
| Prime Minister | Clement Attlee |
| Chancellor of Exchequer | H. Dalton |
| | Sir S. Cripps (13 November 1947) |
| | H. Gaitskell (19 October 1950) |
| Board of Trade | Sir S. Cripps |
| | H. Wilson (29 September 1947) |
| | Sir H. Shawcross (24 April 1951) |
| Ministry Agriculture and Fisheries | T. Williams |
| Ministry of Labour and Nat. Service | G. Isaacs |
| | A. Bevan (17 January 1951) |
| | A. Robens (24 April 1951) |
| Ministry of Transport | A. Barnes |
| Ministry of Food | Sir B. Smith |
| | J. Strachey (27 May 1946) |
| | M. Webb (28 May 1950) |
| Ministry of Fuel and Power | E. Shinwell |
| | H. Gaitskell (7 October 1947) |
| | P. Noel-Baker (28 February 1950) |
| Ministry of Supply | J. Wilmot |
| | G. Strauss (7 October 1947) |
| Ministry of Economic Affairs (combined with Exchequer 13 November 1947) | Sir S. Cripps (29 Sept.-13 Nov. 1947) |
| | H. Gaitskell (28 Feb.-19 Oct. 1950) |

# Index